THE
SCIENTIFIC
ATTITUDE

About the Book and Author

Science is many things: a way of thinking and an activity of individuals in the laboratory; a highly structured institution that recruits, instructs, and regulates its members; a sensitive, interactive, and integrated segment of modern culture and society. Professor Frederick Grinnell presents in this valuable text and survey a clear and comprehensive introduction to all these aspects of science from the point of view of the scientist.

Using many examples, drawn primarily from the biomedical sciences but also from everyday life, the author provides the ideal general introduction to science studies. He goes beyond narrow considerations of methodology to consider broader questions of science as attitude, process, institution, and social force. The text is enhanced by the author's familiarity with several philosophical traditions and the light they throw on the scientific attitude.

The text is straightforward, free of jargon, and completely accessible to beginning students as well as to scientists and laypersons. Professional scholars will also profit from the book's unique blending of the various perspectives on science.

Frederick Grinnell is professor of cell biology and anatomy at the University of Texas Health Science Center at Dallas.

THE
SCIENTIFIC
ATTITUDE

FREDERICK GRINNELL

WESTVIEW PRESS

BOULDER & LONDON

Copyright © 1987 by Westview Press, Inc.

Published in 1987 in the United States of America by Westview Press, Inc.; Frederick A. Praeger, Publisher; 5500 Central Avenue, Boulder, Colorado 80301

Library of Congress Cataloging-in-Publication Data
Grinnell, Frederick.
  The scientific attitude.
  Bibliography: p.
  Includes index.
  1. Science—Philosophy.   2. Science—Methodology.
I. Title.
Q175.G754 1987   501   87-10107
ISBN 0-8133-0539-X
ISBN 0-8133-0540-3 (pbk.)

Printed and bound in the United States of America

∞ The paper used in this publication meets the requirements of the American National Standard for Permanence of Paper for Printed Library Materials Z39.48-1984.

10   9   8   7   6   5   4   3   2   1

*For Paula, Laura,
Phillip, and Aviva*

Let us learn to dream, gentlemen; then perhaps we will find the truth. But let us beware of publishing our dreams until they have been tested by the waking understanding.
—August Kekulé von Stradonitz

Contents

Illustrations

Tables

Figures

Preface

My 7th-grade science teacher, Leon Perkins, characterized science as "serious play." That sounded good to me, and, after completing the typical secondary school science curriculum, I went on to major in chemistry in college. By the time I reached graduate school, I was committed to what Sir Peter Medawar has called the "myth of scientific induction."[1] Popper and Kuhn were unknown to me, and my knowledge of philosophy of science consisted of a series of discussions I had engaged in some years earlier with a sociologist named Richard Rehberg.

There were features of science, however, that I found puzzling and bothersome. An example: I knew that science could be used to control natural events very precisely. But when it came to testing hypotheses, it seemed to me that we were taught to be satisfied if we could predict experimental results "to a first approximation." Also, once I began doing research myself, it soon became clear that science sometimes gets accomplished in unpredictable ways. For instance, in graduate school I studied with an enzymologist named Jon Nishimura who was investigating, among other things, the catalytic mechanism of the enzyme *succinyl coenzyme A synthetase*. Jon gave me a file of papers to read, one of which described preparation of a chemical analog of *coenzyme A*. I assumed that he included the methods paper because he wanted me to use the technique. Therefore, I prepared the analog, and we studied its interactions with the enzyme. This approach turned out to be very successful, and I told Jon how impressed I was with his foresight about the usefulness of the analog. He, however, was under the impression that I had come up with the idea.

As a consequence of my experiences, I began to suspect that there were many scientific methods, not just one Scientific Method, and I became interested in the philosophy and sociology of science. Like most other scientists, I had never studied the meaning of science in a philosophical or sociological sense. As students, we take formal science courses, do extensive laboratory work, and usually are awarded a degree for technical competence. We are skilled apprentices in an art and learn its underlying assumptions in a way that meets our practical needs rather than in a systematic fashion. But, in the process of training students and postdoctoral fellows, I have found that it is very helpful for young investigators to think about the assumptions of science in an explicit manner. It improves their research and helps them understand how they fit into the larger scheme of things both inside and outside of science.

Most books written in the traditions of philosophy and sociology of science are difficult for scientists to read and understand. Partly, this is a result of the specialized jargon of these fields. Partly, it is because philosophers and sociologists are observers who usually view science and scientific discovery retrospectively. Scientists, on the other hand, experience themselves as being engaged in the process of discovery. When the theory of science is presented separate from the day-to-day practice of science, it becomes too abstract for the pragmatically minded investigator.

I decided to write a book about science that would introduce important issues in philosophy and sociology of science without getting into the fine details and current controversies of these fields. Rather, the primary and systematic focus of this work is engagement in science, which is what I know best; theoretical and practical issues are raised secondarily and eclectically in this context. Also, I have written this book so that it can be read at different levels. For undergraduate students (and nonprofessionals interested in science) it is a broad picture of how science is done and the implications of doing science for humankind; for graduate students it is an introduction to the collective thought processes and politics of science that should make it easier to survive in graduate school and beyond; and for professionals it is an invitation to reflect on the conduct and direction of modern science.

The book is organized into seven chapters. Chapter 1 introduces the general approach, an inquiry into the assumptions that scientists take for granted as they are doing science. Chapters 2 and 3 deal with the activities typical of individual scientists: making observations, designing experiments, developing hypotheses. Chapters 4 and 5 concern the collective aspects of science: transmission of the assumptions of science through the training process, and maintenance of the assumptions of science through collective evaluation of an individual investigator's work. Chapter 6 contains a discussion of how science relates to other activities in which people are engaged, especially religion and politics. Finally, the conclusions in Chapter 7 reiterate the major points made earlier.

Many of the examples I have used come from the biomedical sciences. I have chosen these examples because they are part of the research

research fields with which I am most familiar. Otherwise, I would be writing about others' experiences instead of my own. But for me, science is not a spectator sport. My hope is that, by relating the examples I have chosen to general conceptual issues, I will have made it possible for readers to draw parallels from their own fields.

Frederick Grinnell

Acknowledgments

In preparing this book I had a great deal of help. My friend and colleague William Snell constantly was ready to discuss and help clarify the important issues. David Hull helped me through the first two revisions of the manuscript. Others I wish to thank for their comments and suggestions are Richard Anderson, William Bechtel, Rupert Billingham, William Curtis, Steve Fuller, Patrick Heelan, Gary Hunnicutt, Benjamin Lewin, John Peeler, John Pfeiffer, Karen Reeds, Jay Tashiro, Quentin Wheeler, and Jean Wilson. Finally, I am indebted to David Hausman and Richard Zaner, faculty members in the philosophy department at Southern Methodist University, who were willing to help an assistant professor of cell biology from the medical school across town begin learning about philosophy of science; and to Maurice Natanson, professor of philosophy at Yale University, with whom I had the opportunity to study for a semester in 1981 when I was on sabbatical, and who later convinced me to write this book.

F.G.

Introduction

1 What do scientists do when they do science? According to the advertisement at the beginning of "The Double Helix," J. D. Watson's account of how the structure of DNA was discovered, doing science includes "politics, sex, wine, movies, teamwork, rivalry, genius, stupidity, and virtually everything else that makes life in the lab and out something less than perfect and a great deal more than dull."[1] "A great deal more than dull" is an understatement. As Lewis Thomas puts it:

> Scientists at work have the look of creatures following genetic instructions; they seem to be under the influence of a deeply placed human instinct. They are, despite their efforts at dignity, rather like young animals engaged in savage play. When they are near to an answer their hair stands on end, they sweat, they are awash in their own adrenalin. To grab the answer, and grab it first, is for them a more powerful drive than feeding or breeding or protecting themselves against the elements.[2]

In this book, I have focused on the activities of scientists. My purpose is to try to offer an understanding of science as one of the many types of activities in which people are engaged. Some activities, such as eating and sleeping, are typical of all people. Other activities are typical of people acting as detectives—for instance, soliciting information and gathering evidence to reconstruct the events associated with a crime. When people act as scientists, they are engaged in a special kind of detective work that includes gathering evidence about the regular features of observed natural events and imagining hypotheses to explain these regularities.

Several questions immediately arise. For instance, are there criteria that make some activities "scientific" and others "nonscientific"? How does one learn how to carry out scientific activities or, more precisely, how to carry out activities scientifically? An essential distinction must be made between the "scientific method" as an abstract concept and science as a sometimes rewarding, sometimes frustrating, always challenging process in which people are engaged.

Philosophers of science often reflect on the abstract features of science and leave to sociologists and psychologists the practical problem of how science is carried out.[3] Nevertheless, it has only been by studying science as performed by particular scientists that philosophers have come to understand what they think constitutes the "scientific method."

According to an old story, there are three types of baseball umpires: The first says: "I call them [balls and strikes] as they are"; the second says: "I call them as I see them"; and the third says: "What I call them is what they become." It is a common error to think that scientists are like the first type of umpire. Although calling them as they are is the scientist's goal, what scientists do is to call them as they see them, and what they call them often is what they become. What constitutes acceptable science at any particular time in history is determined by the collective assumptions of working scientists. The problem, then, is to understand what the meaning of science is—not abstractly, but in terms of how scientists routinely do science.[4]

It is useful to think of science as a serious game played according to rules that are agreed to by scientists. These rules concern not only what counts as evidence but also the permissible ways in which the evidence can be presented. As in every game, both these aspects must be taken into consideration if one is to understand how the game is played. In poker, for instance, the cards count as evidence, but the best cards do not always win: Against a successful bluff, the best cards may be folded before the hand ends.

The ability of modern science to explain and predict natural events has led some philosophers to the view that the world can be understood only through the application of scientific ideas and methods. Others, however, have argued against this notion, pointing out that the ideas and methods of science originate in our experience of daily life, which we all take for granted.[5] One becomes a scientist by viewing the world in a particular manner;[6] scientists select for study those aspects of the world that are amenable to analysis by scientific methodology. A person acting as a scientist constructs a scientific domain out of the world when s/he adopts a scientific attitude. But people also construct such additional domains of the world as religion, politics, art, and poetry by application of other attitudes.[7] Scientific, poetic, and religious attitudes about the sun, for instance, can be illustrated easily by the expressions below:

All the spheres revolve about the sun as their midpoint, and therefore the sun is the center of the universe.

—Copernicus

But, soft! What light through yonder window breaks. It is the East and Juliet is the sun.

—Shakespeare

And God made the two great lights: the greater light to rule the day, and the lesser light to rule the night.

—*Genesis*

Typically, people who at one time carry out activities that are scientific, at other times engage in a variety of activities outside of science. And it becomes difficult to separate the scientific activities of scientists from influences outside of science. Consequently, in order to develop a complete view of how science works, one must analyze science both as an individual activity and in relation to other activities of daily life.

Consider the scientific activity of selecting particular research problems for study. Interestingly, two investigators often cannot agree on whether a particular research problem is worthwhile studying or whether the studies that already have been carried out constitute a satisfactory solution to a problem. Intellectual confrontation is part of science. Two questions, therefore, must be addressed regarding the selection of research problems. First, what is it that distinguishes a problem in need of a solution from a problem that already has been solved? Second, why do scientists choose to study some problems and not others?

In concrete terms, investigators often choose to study particular problems because the research will be fundable (by a granting agency), do-able (with currently available methodology and resources), significant (with respect to current problems in a given field, such as biology), or intrinsically appealing (to the investigator's curiosity). Each of these criteria suggests a variety of additional concerns. Who decides what makes a problem fundable and on what basis? In what ways do available methodologies control the scope of scientific investigations? How are current problems in biology identified? And, finally, what makes a particular problem intrinsically interesting, and why at one time but not another?

In short, a variety of factors enter into the identification and selection of research problems. Some of these factors might be considered scientific, others turn out to be political, and still others are religious. In any case, since progress in science requires that investigators identify and select new research problems, an analysis of the factors involved in selecting problems for study is required to understand how science works. In taking the above approach I have adopted an unusual way of thinking about the activity of selecting research problems. Typically, one takes it for granted that selecting research problems is a routine activity of scientists and gives the matter no further consideration. In the above analysis, however, I have stepped back from this routine activity and reflected on its meaning and the assumptions that scientists make in carrying it out.

This method of analyzing the underlying assumptions of human activities is a useful way to learn what the activities mean to the people who are performing them.[8] Consider, for example, the problem of what constitutes a work of art. A freshly painted wall in a house is rarely considered a work of art. A large textured canvas painted white, however, might be found hanging in a museum. (There is such a work, but painted all black, at the Art Institute of Chicago.) In principle, by analyzing what a group of people takes for granted to be art, one could arrive at an understanding of what art means to those people. In the subsequent chapters of this book, I have tried to analyze what scientists take for granted to be science, and in so doing I have tried to reach an understanding of the meaning of science and how it works.

Making Observations

Eskimos can observe thirty different kinds of snow and ten different kinds of mud, and their survival often is dependent on recognizing these differences. On the other hand, they notice only one type of summer flower, or at least they have only one word for summer flower.[1] Similarly, gauchos (South American cowboys) recognize two hundred different horse colors, but divide the plant world into four "species": fodder, bedding, woody materials, and all others.[2]

Examples such as these point up the fact that there are significant differences in what people experience when they look around themselves. If several people erroneously think that they are experiencing the same thing, then they can become as confused as the blind travelers who, according to an Indian fable, came upon an elephant in the forest. "Get back, it's a snake," says the first, who touches the trunk. "No! It's a rope," says the second, who grabs the tail. "I think it's a tree trunk," says the third, who encounters a leg. In science, it is essential to avoid such confusion as much as possible. This is why it is so important to understand what scientists mean when they talk and write about the observations they have made.

In this chapter, I will discuss what an investigator means when s/he observes and thinks about objects. The word *object* here is used in a broad sense that includes not only things but also events, processes, and situations. Emphasis is placed on the importance and influence of an observer's preexisting concepts on the observations that are made. For an individual scientist, the challenge and excitement of the game of science begins with learning how to see new things or how to understand old things in new ways. By reflecting in the following sections on the current concept of "cell," the reader will reach an understanding of what biologists have in mind when they talk about biological organisms—a consideration that is pertinent to understanding the notion "alive." Finally, I will suggest that the impartiality of science resides in its collective aspect.

OBSERVING CELLS

Many biologists spend their time studying cell structure and function. To carry out such studies, these investigators must make many assumptions; one central to the entire enterprise is that cells are objects that can be observed and thought about. It is fair to ask, therefore, what one has in mind when thinking about or observing cells.

I can look up *cell* in a textbook[3] and learn that the term was first applied by Hooke in 1665 to the individual chambers observed in cork—chambers now known to be empty spaces where cells once resided. From this beginning, Virchow's (1858) cell theory eventually developed: Living things are composed of one or more units called cells, which arise only from other cells, and cells are capable of maintaining their vitality independent of one another. Also, one can look up the word *cell* in a dictionary and read that a cell is a "usually microscopic plant or animal structure containing nuclear and cytoplasmic material enclosed by a semipermeable membrane, and, in plants, a cell wall."[4]

The historical development of the cell concept is interesting and instructive, and the dictionary definition points to some important features of cells. Nevertheless, these considerations fail to resolve the question of precisely what the practicing investigator means when s/he looks at something in the microscope and says, "That's a cell." More explicitly, what is it that the investigator has in mind that results in the identification of some objects as cells and other objects as something else, but not cells?

There is a difference between abstract ideas of things and experiences of specific examples of things.[5] Scientists never experience an "ideal" cell. Rather, they observe specific examples of cells. Figure 2.1 shows several examples of somatic cells in human tissues as they are observed and photographed at the same magnification through the light microscope. The cells show a remarkable degree of diversity in size and composition. Red blood cells (R) and white blood cells (W) can be seen inside a capillary in Figure 2.1A. The red blood cells lack nuclei and contain very little in their cytoplasm except the protein hemoglobin. A row of epithelial cells (E) lining the trachea can be seen in Figure 2.1B. These cells have fine extensions called cilia (c) at their apical surfaces, and nuclei (nuc) are located in the basal ends of the cells. Several skeletal muscle cells (S) are shown in Figure 2.1C. Each cell is very long and contains many nuclei (nuc). The lines in the cytoplasm of these cells reflect a special arrangement of contractile proteins. Finally,

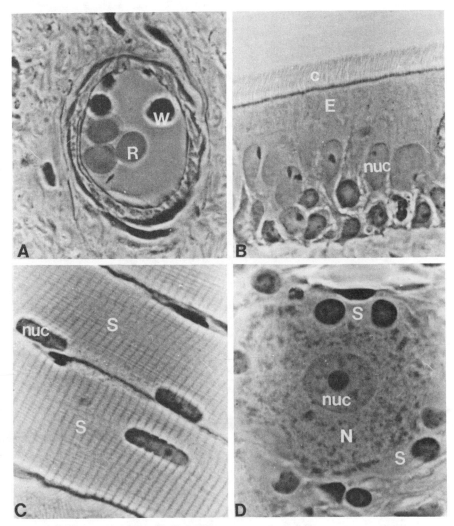

Figure 2.1   Somatic cells in human tissues (x1150)

the body of a nerve cell (N) surrounded by satellite cells (S) is shown in Figure 2.1D. The nucleus of the nerve cell body (nuc) contains a prominent nucleolus.

After looking at the examples in Figure 2.1, an untrained observer might wonder how I know that these objects are "typical" examples of cells. Somehow, out of my experience of looking at and studying particular examples of cells, I have built up a practical concept of what constitutes a cell. Having built up this concept, I use it when

looking at and evaluating other particular examples of objects that
might be cells.

ANALYZING THE CELL CONCEPT

Usually, investigators employ the practical concept of a cell that they
have acquired without worrying about its content. Nevertheless, I can
choose to reflect on my cell concept and thereby determine the structural
and functional features that I have in mind when looking at or thinking
about cells. The method I use to accomplish this is to imagine variations
in the concept.[6] According to this method, I should begin by thinking
of a typical example of a cell. Then I should imagine variations in
the example: no nucleus, no plasma membrane, no proteins, no nucleic
acids, and so on. Some of these imaginary variations will still fit into
my practical concept of a cell; others will not. By carrying out this
analysis systematically, I will end up with a reflective awareness of a
set of functional and structural boundary conditions. Anything within
the range encompassed by these boundaries is what I call a cell; anything
outside those boundaries is something else.

Because the method of imaginary variations is unusual, it is worthwhile to illustrate how it works by considering a familiar object of
daily life—a table. Most adults and many children understand what
the word *table* means in its common usage. The definition is partly
functional and partly structural. A table is an object on which one
can place food or play cards. It has a flat surface and four legs. For
most purposes a more precise understanding of what *table* means is
unnecessary. If, however, I wished to be quite precise, I could consider
variations in a typical example of a table to determine if the variations
still fit into the table concept. For instance, a 10 foot by 10 foot flat
surface supported by four 8-foot legs under which a car was parked
might be a carport but probably not a table. A 5 foot by 5 foot flat
surface supported by four 8-foot legs is something else, perhaps a
wedding canopy. In any case, the latter structure is too small to be
a carport but too high to be a table, at least for normal-sized people.
On the basis of such functional and structural considerations, I am
able to define the range of boundary conditions that would allow me
to see an object as a table.

The reason why the table concept can be defined only within a
range of conditions is that "ideal" tables do not exist. One experiences
only real examples of tables. In one's experience of a table, however,
the observer can see only parts of the table. As tables are three-
dimensional, solid objects, one cannot view a table in its entirety at
the same time. Thus, one sees a table from a single perspective, but

assumes that the table could be seen from other points of view and knows how the table would appear from those other points of view. Accordingly, the same table from different perspectives is recognized as different views of the same table, not separate views of different tables. An inexperienced observer who lacks a mature practical concept of a table may be unable, however, to recognize the partial view as an example of a table. Similarly, when a congenitally blind person's vision is restored in adult life, the individual requires a period of time to learn "to see" many typical objects that were previously recognized by touch.[7] In short, one's ability to "see" an object depends on whether the individual already has in mind a concept of what the object looks like.[8]

Returning to the cell concept, one notes that cells are visible objects that can be seen with a light microscope. I cannot imagine a cell that could not be seen. The problem of what it means to see a cell is of major importance and will be discussed. First, however, I will discuss some functional aspects of cells.

Cells can be free-living entities (e.g., bacteria) or may exist as parts of a complex tissue or organism (e.g., skin keratinocytes or liver hepatocytes in mammals). In tissues, cells are interdependent; that is, the normal functioning of different cell types in an organism depends on interactions among the cell types. One can study isolated cells, however, by removing them from their normal situation.

Cells may be able to divide (e.g., skin keratinocytes), or they may have lost the ability to divide (e.g., adult neurons). We can conclude, therefore, that the ability to divide is not an essential feature of cells, only that cells were able to divide at one time. In fact, mature cells do not have to contain nuclei (e.g., human red blood cells).

Finally, cells are subject to the designation of being alive or dead. One typical way of determining whether cells are alive or dead is to mix them with a special dye that stains the cells only if they are dead. Another way is to analyze the medium in which the cells are suspended to determine whether cytoplasmic components have been released from the cells into the medium. Both of these techniques determine whether or not the cells have lost their integrity as a compartment separate from the external environment. That is, cells maintain themselves apart from their external milieu, and the expression of normal cell function requires that this separation be maintained.

The foregoing discussion defines some of the functional boundary conditions of the cell concept. Now for the structural characteristics. Unless one can look in a microscope and distinguish between objects that arc cells and others that are not, one's grasp of the cell concept is incomplete and would be of limited usefulness for carrying out experiments using cells.

LEARNING TO SEE CELLS

Most cells cannot be seen directly by investigators. Rather, the assistance of microscopes or some indirect method is required. One can ask whether there are theoretical differences among objects that can be seen by the natural optical system (one's eyes) and those that require an artificial optical system (a microscope). It would be a serious problem if, by using a microscope, an investigator introduced artifacts into the observations that were made. Significantly, instruments such as microscopes that are used routinely are usually treated by scientists as extensions of themselves.[9] That is, when looking through a microscope, an investigator is concerned with what s/he sees and not with the theory of microscopy or the construction of the microscope.

If one is asked to identify cells under the light microscope but cannot see them, then it is possible that the observer has not yet learned how to use the microscope. As an alternative, one can show an observer a picture of cells and cell debris and ask the question: "Which are the cells?" I have shown such pictures to 1st-grade elementary school students, and they were unable to identify the cells. If, however, I showed them which of the objects were cells and which were debris, the students subsequently were able to recognize other, similar cells. More dramatically, the average medical student often is unable to distinguish between cells and cell nuclei when first shown light micrographs of tissue sections such as those in Figure 2.1. After studying histology the same students are able not only to make such distinctions but also to discriminate among the different tissues and recognize the specific cell types of which they are composed.

The two examples above have some important differences. In the first case the elementary school children had essentially no previous idea of what a cell was but could come to recognize an object as a cell, albeit under very limited circumstances. In the second case, the medical students had a textbook concept of cells; but even so, they were initially unable to "see" particular cells.

The students' inability to see the cells was not a technical problem. There can be technical problems, of course—as when one takes an unstained tissue section and places it under a microscope. Under these conditions it is possible to tell that something is "there," but not precisely what. As discussed in any histology textbook, the reason is that there are few visual features of unstained tissue sections that our eyes can discriminate. As the students were studying stained specimens,

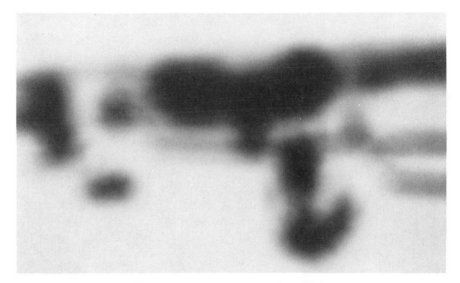

Figure 2.2   Emergent form I

Figure 2.3   Emergent form II

however, sufficient details of the field were observable that could have permitted them to distinguish among different cells and between cells and the noncellular elements of the tissue. Thus, for these students, the cells were visible but unseen.

The elementary students were able at first to identify objects in their

Figure 2.4   Emergent form III (photo by the author)

visual fields and subsequently learned to recognize objects having the particular forms corresponding to cells. This example illustrates the idea that the ability to see a particular object in one's visual field depends not only upon whether the observer can see that "some object is present," an object that stands out from the remainder of the visual field, but also on the observer's ability to recognize the particular form of the object.

To say that some thing exists (Latin: *ex-,* out; *sistere,* to cause to stand) is to say that the thing stands out. The series of photographs starting on page 11 illustrates the distinction between seeing an undefined object present in the visual field and recognizing the particular form of the object. In Figure 2.2, a dark object (or group of objects) stands out from a light background. The identity of the object, however, is obscure. Figure 2.3 presents the same field, but with increased clarity. Now one can see that there are several different objects present, but their forms are still vague. Finally, Figure 2.4 presents the detailed visual field in a clarified form.

It has been suggested that one's ability to see any object is learned. According to this theory, newborn infants do not discriminate among different objects in their visual fields, or they do so only at a very rudimentary level. For the newborn, the distinction between self and nonself is at first vague or nonexistent. During the first two years of life, however, the external world becomes full of objects as the infant becomes aware of self in relation to nonself. This so-called intellectual revolution results in the infant's ability to use concepts practically—concepts such as object, space, and time.[10]

It may be difficult for a fully aware adult to imagine the various stages in an infant's developmental recognition of self versus nonself. A helpful analogy is the experience of a blind person learning how to use a cane. At first, the person feels only the cane against the hand. The cane is experienced as nonself. Eventually, however, the cane is experienced as an extension of the hand. Then, the tip of the cane can be used to feel for the presence of objects. Finally, different types of objects can be recognized as a mental picture of the environment if built up.[11]

In order to understand the infant's intellectual revolution, it is necessary to realize that one's conscious acts have certain typical features. Usually, these features are implicit unless one reflects on the act of experiencing. For instance, in my experiences of the world I am aware of myself, as opposed to some stranger, as the one who is doing the experiencing. I am aware of my spatial location. "Here" is where I am, not over there or around the corner. And I am aware of my temporal location. The time when I am having this experience is "now," not earlier, not later. Necessary to my observation of an external object is my ability to recognize that it is "not part of me." I can distinguish self from nonself even though we share the same "now." It is "there," and I am "here." Finally, I expect the same object to have the same appearance from one moment to the next. Things do not appear and disappear randomly.[12]

To put it another way, the infant has at first an undifferentiated field of perception, what William James described as a "big blooming buzzing confusion."[13] Within this field, the individual eventually comes to recognize specific types of objects and to distinguish self from nonself. The operational "concepts" by which the infant experiences the world have been given the technical name *schema* (sing. *schemata*), and the world is experienced according to how well perceptions fit into the infant's schema.[14] For instance, sucking is initially a reflex response to anything placed into an infant's mouth. With time, however, milk producing and non-milk producing objects can be distinguished. The original sucking schemata was used to assimilate new information. The infant accommodated to this new information by developing two closely related but distinct schema that now allow the infant to experience the world in a more discriminating manner. In short, schema are practical concepts in flux. The dialectical process of assimilation of new experiences by schema and their accommodation to the new experiences is a developmental process that results in an increasing sophistication

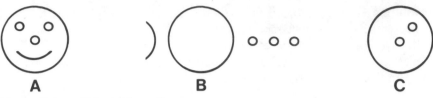

A   **B**   **C**

Figure 2.5   Schematic outlines

of these concepts. As will later be discussed, scientific hypotheses develop
in the same manner.

THE CELL GESTALT

In addition to seeing an object as standing out, an observer's ability
to recognize an object as "one of these" depends on being able to
recognize the form of the object. Most observers recognize Figure 2.5A
as a face. Yet the figure is highly schematic and not like any "real"
face. Nevertheless, one can see the schematic outline of the face, the
eyes, the nose, and the mouth. It is essential, however, that the features
be in a particular spatial arrangement. Figure 2.5B, for instance, has
the same elements as 2.5A, but they are arranged differently and the
face cannot be visualized. An observer, especially an astronomer, might
see a diagram of the sun and its planets. Finally, one can remove the
mouth and one eye, and what remains are two circles enclosed in a
larger circle. The face has disappeared altogether.

The form of the idealized structure in Figure 2.5 is known as a
gestalt. Gestalts are unified wholes with the characteristic that the
arrangement of component features provides information not apparent
in the features alone.[15] This becomes clear when we compare Figures
2.5A and 2.5B, where the same components are arranged differently.
In addition, there is a reciprocal relationship between the components
and the intact structure. *Only by seeing the individual features of the
face can one see the face as a whole, and only by seeing the face as a
whole can one give meaning to its individual features.* Significantly, the
boundary conditions of a particular gestalt can be determined by using
the method of imaginary variations already described for cells and
tables. For instance, by making imaginary variations in Figure 2.5A,
one learns that the face still can be seen if either the outline of the
face, one eye, the nose, or the mouth is deleted. The face no longer
can be seen, however, once the mouth and one eye are removed.

Returning to our discussion of the concept of cell, it is evident that
visual images of cells—like visual images of faces—have gestalt features.
Although one takes it for granted that cells exist, the cells can be seen

only if the observer is already familiar with cell form. The elementary students began with the concept and ability to recognize objects-in-general, but not cells as a particular kind of object. Once the students were shown an example of cell form, however, they assimilated the information, and their preexisting object concept accommodated to that information in such a way that the particular object form of cells could be recognized on subsequent occasions. The same events occurred with the medical students, but the medical students went much further in that they learned to recognize not only the forms of cells-in-general but also the forms of particular types of cells.

The above considerations lead to the conclusion that the idealized objects of science have the features of gestalts, and that each type of practical concept of an object is characterized by interrelated structural and functional features. An observer can see and think about individual objects as examples of a particular type only if the features of the individual object appear to be within the boundaries established by the preexisting gestalt with which the observer is familiar. Otherwise the object in question cannot be recognized: "I've never seen one before."

Observing a particular type of object presupposes that the concept of the object is already in the observer's mind; in addition, the presence of the concept prevents the observer from seeing things in other ways. The lack of detail in Figures 2.2 and 2.3 allows the observer to imagine the presence of many different novel images in the field. Having seen Figure 2.4, however, the observer would find it very difficult to return to the earlier figures and see anything other than Figure 2.4 defocused. This illustrates the crucial point that *seeing things one way means not seeing them another way.*[16] This feature is a difficult block that must be overcome by scientists if practical concepts are to be modified so as to assimilate newly discovered information.

Once we understand that the observation of biological objects depends upon prior acquisition of particular concepts, it becomes clear why individual investigators cannot be impartial in an ideal sense. Each person's previous experiences will have led to the development of particular concepts of things, which will influence what objects can be seen and what they will appear to be. As a consequence, it is not unusual for two investigators to disagree about their observations if the investigators are looking at the data according to different conceptual frameworks. Resolution of such conflicts requires that the investigators clarify for each other the concepts that they have in mind.

When I started working on cell adhesion in 1970, I found that different investigators meant different things when they used the word *adhesion.* Some meant the attachment of cells to each other; others meant the attachment of cells to extracellular surfaces. Within the latter group of scientists, some investigators studied the numbers of cells that were able to attach to surfaces; others were concerned with the shape of cells after they attached to surfaces; and still others analyzed the forces necessary to remove cells from surfaces to which they had been previously attached. But everyone used the same word—*adhesion*—as if they were all studying the same process. Consequently, a participant could hear one investigator at a scientific meeting report that cell adhesion requires energy and hear another report that cell adhesion does not require energy. It took me years to develop a mental program that included the players (scientists) and their positions (what they meant by adhesion) so that I could make sense of the dialogue in the field.

Such differences in meaning may be very subtle, but they can have major consequences. For instance, during the fight to eradicate smallpox in Bangladesh, the goal was to vaccinate everyone who came into contact with infected patients. Patients were asked to provide the immunization teams with lists of visitors, but outbreaks continued to occur despite the massive immunization effort. Then, one astute health worker learned that patients never included the names of relatives on their lists, and he finally realized that *relatives were not considered visitors.*[17]

In understanding the important role of preexisting gestalts in the ability of scientists to make observations, one can gain some insight into the permanently tentative nature of science.[18] Historically, scientists have often changed their conceptual frameworks and ways of looking at the world. Consequently, what once was thought to be true later was considered a falsehood or only an approximation, while previously discarded notions reemerged and were accepted as valid. The practical concepts that permit observations to be made are only limited approximations. As such, they are subject to continual change. Like schema, they accommodate to new information that is assimilated.

During the several hundred years since the time of Robert Hooke, the "scientific" view of cells, defined by the prevailing cell concept at any particular time, has changed dramatically. As long as cells were thought of as hollow spaces within tissues, they could not be thought of as the basic building blocks of which tissues were composed. As long as the production of progeny cells during cell division was believed to involve direct reproduction of all the cell contents, the role of the nucleus and genetic material in cell division could not be discovered. And as long as the metabolic activities of cells were thought to depend on vitalistic forces, the biochemical basis of metabolism and the role of enzymes as chemical catalysts could not be elucidated.[19] Presumably, whatever the "real" nature of individual cells, it has remained the same over the past several hundred years. Nevertheless, as scientists' concepts of cells have changed, what scientists have seen when they have looked at cells also has changed.

For an individual scientist, much of the challenge and excitement of science is learning how to see and understand things in novel ways. A good example is the development of ideas about genes. For a long time after Gregor Mendel's discoveries, genes were considered formal units that could be used to describe the mathematical relationships of inheritance. It was not until 1940 that a biochemical relationship between genes and enzyme expression was shown by George Beadle and Edward Tatum. Four years later, Oswald Avery, Colin MacLeod, and Maclyn McCarty were studying pneumococcal bacilli (their purpose was to develop a vaccine as part of the war effort) and discovered DNA transformation.[20] Most scientists, however, were unable to think of DNA as genetic material. Few previous connections between bacteria and genetics had been established; bacteria were thought not to have chromosomes; and the chemical structure of DNA appeared to be too simple to explain the information content of genes.[21] Some investigators, however, were able to see things differently. In "The Double Helix,"[22] Watson described why it came about that Salvador Luria arranged for him to go to Europe to study the chemistry of DNA. Watson says that "Avery's experiment made it [DNA] smell like the essential genetic material." The situation is like listening to a series of musical notes and suddenly recognizing the tune when one additional note is heard. At that moment, all of the previously heard notes take on new significance and meaning.[23]

Despite the amazing advances in molecular biology since 1940, the relationship between the genotype (hereditary makeup) and phenotype (visible traits) of an organism remains poorly understood. Although distinctions can be made between almost every bone in humans and chimpanzees, the proteins of these two species have almost identical amino acid compositions. Moreover, humans have only one more chromosome pair than chimpanzees. On the other hand, different species of salamanders in the genus *Plethodon* are difficult to distinguish visually. Yet their genetic compositions are much more different than the differences between humans and chimpanzees.[24] The solution to the problem of the relationship between genotype and phenotype may require a new gestalt of the gene concept. Perhaps the gene concept should include not only the DNA present in germ cells but also the nature and/or topographical location of other types of molecules within these cells.[25]

THE CONCEPT OF LIFE

Returning to the functional features of the cell concept, we will find it worthwhile to consider in more detail the notion that cells can be alive or dead. Earlier, I suggested that this distinction depended on the idea that cells maintain themselves separate from their environment. Most cells are surrounded by some sort of membrane or envelope, and interactions between the cells and their environment occur first at the level of this membrane. For cells to exist as cells, these interactions are critical. It is from the environment that cells take up the nutrients required for the maintenance of cell structure and function, and into the environment that cells secrete many of their products. Implicit within the cell gestalt, therefore, is the assumption that a cell is part of a system—that is, a smaller compartment (the cell) within a larger compartment (the surrounding environment). The two compartments interact with chemical information flowing in both directions.[26]

Thermodynamically speaking, individual cells are not in self-equilibrium since a constant exchange of biochemical substances occurs across the cell's boundaries. This exchange is required because the functional integrity and identity of each cell require that it continue to maintain itself separate from the adjacent or surrounding compartments. Accordingly, biologists typically are not interested in treating cells as objects of inquiry in the same idealistic way sometimes associated with physical science. In thermodynamics, for instance, it is possible to treat objects as ideal entities within an isolated, closed system (e.g., gas molecules as particles in a box). The boundaries of the isolated system separate the system from its surroundings, and what defines the system as being closed is the lack of exchange of matter (energy) across the boundaries.[27] *In the context of biology, the meaning of objects such as cells includes their surroundings.*

The current concept of biological organisms assumes the establishment and maintenance of biological interfaces to be a necessary and fundamental feature that makes the existence of the organisms possible. The presence of an active interface of exchange between biological compartments such as organisms and their surroundings is an indication that the compartment (organism) under study is alive. At one end of the biological scale, in the research on the chemical origins of life, the goal is to find the conditions under which a self-replicating system develops and maintains itself separate from its chemical environment. At the other end of the scale, the problem with finding a physiological definition of death can be understood as a problem of how to assess the presence of an active interface between an individual person and the surrounding environment. For instance, a Harvard Medical School committee concluded that irreversible coma (brain death) could be established by the following criteria: (1) lack of response to external stimuli or to internal need, (2) absence of movement or breathing as observed by physicians over a period of at least one hour, and (3) absence of elicitable reflexes.[28]

THE IMPARTIALITY OF SCIENCE

The final topic to be discussed in this chapter is the problem of the impartiality of science and scientific observations. That science is impartial is a basic assumption made by most scientists. An important clue to understanding this assumption is my awareness that others seem to be able to recognize the same objects of science that I can recognize. My students learn a practical cell concept that is very similar to the concept with which I am familiar. We can talk about cells together, and then we can look at the cells together in a double-headed microscope. Indeed, it seems to me that I have transmitted my cell concept to the students.

The belief that I can teach students how to think about and see cells involves two assumptions. First, I assume that cells are real objects. Second, I assume that students are *other persons who can see and think about things the way that I do.* The latter idea is the intersubjective feature of science. Here, intersubjectivity refers to my recognition of others as people who, in some fundamental ways, are like me—and, in particular, to my recognition that others' basic experience of reality is similar to mine. If they were standing where I am standing, they would see something very similar to what I see.[29] As I will later discuss in detail, scientists act as if the scientific enterprise were universally intersubjective, as if the observations made by one scientist could have been made by anyone and everyone. This notion is what leads scientists to believe that their observations are impartial. In other words, I assume that my observations are not a result of my personal biases since I believe that they can (potentially) be verified by everyone else.

Two very important consequences result from the expectations of scientists regarding the possibility of intersubjective verification. First, only objects of experience that *potentially* are available to observation by *all* suitably trained investigators can be subjects for experiments. This is the reason that the scientific attitude presents only a limited view of the world. What will be emphasized in Chapter 6 is that some aspects of experience described by other attitudes (e.g., religious or poetic) cannot be analyzed through scientific inquiry.

Second, although intersubjective verification *makes possible* the impartiality of science, it does not *guarantee* that scientists will be impartial. Perhaps it would be more accurate to understand impartiality as a goal rather than as an accomplishment of science. This view would be open to recognition of the collective biases held by investigators. These biases, which change from time to time, are the normative attitudes of scientists. As normative attitudes change, previous results no longer are viewed in quite the same way.

Experimental Design and Interpretation

3 If you put a bottle of wine in the freezer to chill it quickly and then forget to take it out, there may be pieces of glass and clumps of frozen wine all over the freezer the next morning. You have made an unexpected observation. Maybe the bottle was cracked. The next time it happens you get suspicious. Recalling from high school science that water expands when frozen, it occurs to you that the wine expanded during freezing, and perhaps the bottle was not large enough to accommodate the expansion. You have proposed a hypothesis that can explain what happened. The bottle broke because the wine expanded when it froze. Moreover, this hypothesis permits you to make some predictions: For example, if you put a full bottle of wine in the freezer, the bottle will crack; if you put a half-full bottle of wine in the freezer, the wine will freeze, but the bottle will not crack. You are ready to begin experiments to test your hypothesis.

In Chapter 2, I discussed the observations that scientists make. Sometimes such observations are unexpected and occur as a part of routine experience, as in the wine bottle incident. Usually, however, scientists make observations as part of experiments that are designed to solve a problem. In this chapter, I will focus on the ways in which an investigator's interests and previously held beliefs influence the research problems that s/he selects for study, the kinds of experiments that are carried out to try to solve the problems, and the ways in which the results of the experiments are interpreted. Of particular importance is the recognition that the adequacy of experimental design depends on both explicit and implicit assumptions made by an investigator. In this chapter, I will also focus on the natural evolution of hypotheses from initial proposition to acceptance to modification or rejection. Finally, I will consider the importance of studying biological systems at different levels of organization.

CHOOSING PROBLEMS FOR STUDY

One can imagine a scene in the woods where an individual who is in the process of chopping down trees is observed by another. Assuming that the observer has previously acquired the necessary practical concepts to recognize that "there is another person who is chopping down trees," it is possible for the observer to formulate a number of different hypotheses about the events going on in the scene. Some of these may relate to the trees: "Those are maple trees," or "there appears to have been a fire in this woods recently." Others may relate to the act of chopping: "That ax must be sharp," or "chopping down trees requires the activity of a particular group of muscles." Still others may relate to the individual who is doing the chopping: "Chopping wood is probably the way that he makes his living," or "he probably heats his home with a wood-burning stove." In any event, the observer can look at the scene in a multitude of ways, each of which involves different hypotheses regarding what s/he sees. Therefore, it has been suggested that an observer imposes a particular meaning on a scene according to his/her interest and interpretation of what is going on.[1] This idea is the basis for the party game in which numerous people are shown the same picture and asked to write down individually what they are looking at. The point to be emphasized is that, in large part, *an observer's previous knowledge and experience determine what aspects of a scene will be interesting to the observer.*

The influence of a scientist's previous knowledge and experience on how s/he does research can be illustrated by an analysis of the many different activities necessary for performing a research study. These include (1) choosing a problem to study, (2) choosing a test system that can be used to study the problem, (3) designing specific experiments to manipulate the system, (4) implementing the experiments, (5) recording the results, and (6) analyzing and interpreting the results. These activities are carried out in the context of the overall research interests in a laboratory, which can be described in terms of three different but interelated elements: (1) the long-term goal, (2) the specific aims by which the goal will be achieved, and (3) the significance of the goal.

The long-term goal is the general problem under study in the laboratory, such as determining the mechanism by which normal cells become cancerous. Implicit in the long-term goal is the investigator's attitude toward the goal as a problem in need of a solution. The investigator is aware of the state of the field based upon knowledge gained from taking science courses, reading published reports, attending professional meetings, and holding private conversations with other investigators. According to prior knowledge, the investigator will think of particular aspects of the field as being explained by well-established hypotheses and not in need of further study, explained by hypotheses that require a little more work to firm up, or problematic and in need of clarification.

Even if a problem is thought of as requiring further study, it will not be selected for study unless the investigator views the problem as being amenable to study. S/he must be able to imagine hypotheses that are relevant to the long-term goal. Broadly stated, these hypotheses are the specific aims of the research project. They represent possible approaches and solutions to different aspects of the problem. How they are tested depends upon the field of study. In experimental biology, hypotheses are usually tested by means of laboratory studies, but occasionally they are tested through analyses of experiments of nature (e.g., experiments involving individuals with genetic defects). Biologists interested in evolution, on the other hand, often test hypotheses by investigating the natural historical record and, less frequently, by designing laboratory studies. Nevertheless, associated with every hypothesis are methods of analysis that can be thought of as technically feasible (i.e., do-able) and interpretable.

Finally, selection of the long-term goal requires that the investigator view the goal as one worth studying. There are many goals that one could select, but only a limited amount of time and resources are available in any particular laboratory. As a result, the investigator often must choose among several possible long-term goals. This choice reflects the investigator's view of the significance of the goals under consideration.

Significance in the above context can have many different meanings. In a biomedical science such as cell biology, significance often refers directly to health-related problems. The significance of understanding cancer, for instance, may be related to the possibility of devising a cure. Alternatively, problems may be viewed as significant if they concern general structural and functional features of biological systems, such as how cells migrate from one place to another in the organism. Migration is an important cell activity in many different situations such as embryonic development, wound healing, and the invasiveness of cancerous cells. Significance also might refer to development of a specific technique or methodology (e.g., a new kind of microscope or a new method for dating the age of fossil specimens) that could benefit all investigators in the field, independent of the research problems that they are studying. Other important aspects of significance are time and place. Cancer and heart disease are major health problems of the twentieth century, but they were relatively unimportant compared to infectious disease during the nineteenth century. Similarly, tropical diseases are more important health problems in tropical countries than in temperate ones.

The above reasons for selecting particular research problems are scientifically oriented. As an investigator's background is derived from *all* prior experiences, however, decisions regarding scientific problems often are made for reasons unrelated to science. I am told, for instance, that evolutionary biologists often choose to study organisms that inhabit exotic, far-off places. Frequently, investigators select a particular research goal because of career incentives ("this research project is fundable and my promotion to tenure depends upon winning grant support"); because of financial incentives ("these results will lead to a marketable product"); because of a desire for recognition ("whoever solves this problem probably will receive the Nobel Prize"); or because of a personal experience ("Uncle Henry died from diabetes, so I want to cure this disease"). Moreover, there are competitive aspects of science relating to other investigators ("nobody has been able to do this; I want to be the first"). Often, intellectual curiosity is the strongest driving force of all.

Only the scientific elements of research projects—long-term goal, specific aims, and significance—are viewed by the scientific community as appropriate reasons to select a research project. Accordingly, these aspects are described explicitly in grant applications to the National Institutes of Health (NIH) or the National Science Foundation (NSF). Left out of such applications are reasons from outside of science, regardless of the extent to which these reasons influenced the selection of the research projects. Inclusion of such information as part of a research grant application would be viewed questionably, to say the least.

HYPOTHESES AND EXPECTATIONS

Once an overall research problem has been chosen, the day-to-day drama of science resides in the imagination and testing of hypotheses. The historian William Whewell used to call hypotheses "happy guesses." After he became Master of Trinity College, Cambridge, he referred to them as "felicitous strokes of inventive talent."[2] The investigator, possessing incomplete—possibly minimal—knowledge of the problem under study, begins by guessing at hypotheses that can be tested to learn more about the problem. These hypotheses range from being very explicit ("this is the change that I expect to observe if I do such and such") to very vague ("something might happen if I do such and such"). Then s/he designs and implements experiments to test the hypothesis. Subsequently, the results of these experiments are used to formulate new hypotheses, which are tested in turn by new experiments.

A good example to illustrate the research process is the problem of growth regulation of mammalian cells. The specific aims of laboratories engaged in understanding this problem often include characterization of environmental factors that influence cell growth, identification of the cellular mechanisms through which the factors exert their effects, and determination of the physiological situations in which the factors function in the organism. In setting up these specific aims, many assumptions are taken for granted. For instance, it is assumed that cells are directly affected by environmental factors; that enough is known about cell structure and function to make it feasible to determine how environmental factors influence particular cell structures and functions; and that cell growth control plays some function in the normal physiology of organisms.

Individual cells can be isolated from the tissues in which they normally reside and then permitted to grow under laboratory conditions. Usually, the nutrient medium in which the cells are grown contains serum—a complex mixture of proteins and other molecules that remains in solution after blood is clotted. Prior to identification of the specific factors in serum that promoted cell growth, most investigators simply took it for granted that such factors existed.

An investigator interested in isolating the specific component of serum necessary for cell growth might begin with an experiment to make sure the system works as anticipated. That is, cell cultures would be set up in the presence and absence of serum, and cell growth would be compared under these conditions. The prediction would be that only cells cultured in the presence of serum will grow. This is an explicit expectation. Fulfillment of this expectation would confirm the established hypothesis and thereby link the series of experiments with those previously carried out by other investigators.

If the above experiment met the investigator's expectations, then the next step might be to separate the serum into fractions and determine which fractions stimulate growth. The hypothesis in this case would be that there is a serum component responsible for the biological activity. Based on this hypothesis, one might predict that the activity could be found in one of the serum fractions. Having obtained a partially purified preparation, one could then continue to fractionate the serum until a single serum component was isolated that had the growth-promoting activity. Thus, at each step, the investigator will have expectations that are based upon previously performed experiments. And as long as the investigator's expectations are met, the course of the experiments will proceed smoothly.

The results of an experiment or some procedure that is part of an experiment do not always meet an investigator's expectations, however. An investigator who was studying cells that were of monkey origin decided to use monkey serum in the culture medium. This decision presented technical difficulties because monkey serum was not readily available. As there are no commercial sources, the laboratory obtained access to several monkeys whose blood could be collected periodically and used to prepared serum. But this procedure could not frequently be performed on the animals. A more efficient alternative was to collect the animals' blood, remove the cells and return them to the animals, and prepare serum from the remaining plasma. (Plasma is blood from which the cells have been removed.) This method could be performed much more often. Everything seemed quite reasonable until the investigator used the plasma-derived serum in the cell cultures. In contrast to the results using blood-derived serum, the cells did not grow in serum prepared from clotted plasma.[3]

How might an investigator respond to the unexpected observation described above?

1. The investigator might regret changing a routine laboratory method and decide to use only whole blood–derived serum in the future.
2. The investigator might assume that there was a technical error and set up the cell cultures again.
3. The investigator might see the result as an unusual one that should be investigated further.

Implicit in all three of these responses is a recognition of the situation that "things are not working the way that they should." That is, the investigator's expectations have not been fulfilled. The differences in the responses themselves may also be a consequence of the investigator's background, knowledge, and current interests.

THE ASSUMPTION
OF REPRODUCIBILITY

Analysis of the assumptions underlying the above responses is instructive. In the first case, the investigator's response focused on technical failure against the background of the observations made. Why the cells did not grow in plasma-derived serum was not a subject for concern. Implicit in the investigator's response was the assumption that the same technique would produce the same results each time it was used. That is, the technique was assumed to be reproducible.

Investigators know that reproducibility of experiments is a requirement of scientific research. While one can raise questions about unique events, only recurring events can be subjected to scientific investigation. It might be suggested that every experiment with cells is unique since the individual cells under study vary from experiment to experiment. That is, an investigator rarely uses the same cells over and over again. One takes for granted, however, that the individual cells are *typical representatives* of some particular class of cells (e.g., human smooth muscle cells isolated by a particular procedure). What is under study in the experiments are the properties of the cell class, not the individual members of the class.[4] The importance of the individuals is that they allow for the testing of hypotheses about the class.

Once it is realized that science is interested in classes rather than in individuals, several important points can be made. First, as will be discussed later, the scientific attitude applied to people is potentially de-individualizing because it deals with unique individuals—like other subject matter—as typical examples of a class rather than as individuals. Second, there tend to be large differences between the physical and social sciences in the sizes of classes of objects that are under study. A chemist studying the behavior of some molecule may be dealing with 10^{20} (100 million million million) examples of a molecule in a particular study; a cell biologist studying cell behavior may be dealing with 10^6 (1 million) examples; and a psychologist studying some aspect of human behavior may be dealing with 10^2 (100) examples. The smaller the number of individual examples under study, the more difficult it will be to conclude that the behavior of the individuals observed is typical of the class that the individuals are supposed to represent.

Another aspect of the reproducibility assumption concerns one's attitude regarding the relationships between events. A scientist who assumes reproducibility takes for granted that the world experienced in everyday life has an underlying order of necessary relationships among events, and that there are universal laws describing these relationships. Otherwise, there would be no reason to believe that observations made one day would be repeatable the next. Leading to— and perhaps confirming—such a belief is an investigator's previous knowledge of the scientific laws that were useful in explaining or predicting events. Operationally speaking, when one says that an event occurred by accident, the implication is that it could not be explained or predicted.

The distinction between necessary and accidental relationships is inherent in one's everyday experience. Although particular features of the world change in this experience, these appear to be superimposed on a constant underlying structure. For instance, I am still myself even though I am growing older; my house is still my house although it needs to be painted; and so forth. Some of these changes constantly recur, and, as a result, I develop typical expectations regarding the future relationship of one event to the next. Other events occur less often, so I have not figured out how to explain them. My typical expectations are prescientific hypotheses. When I adopt the scientific attitude, I also develop expectations, but in a more systematic way. Now I assume that events are related to one another, and I experiment to determine what conditions control the relationship in question. Moreover, because I also assume that scientific observations are universally valid, I anticipate that others who adopt a similar scientific attitude will be able to notice the events and their relationships as I do.

Returning to the observation that the cells did not grow, it is noteworthy that the response of the second investigator also focused on the technical failure, but in a way different from that of the first investigator. The second investigator assumed that the system was correct, but that a technical mistake had been made. Slight changes in the complex technical manipulations involved in carrying out experiments can lead to different observations. Quite often, when an experiment does not work, it is because of operator error. Consequently, the development of technical expertise often becomes a goal in itself. There is an intrinsic artistic element that is part of laboratory manipulations, and many investigators develop reputations for their particular skills.

SEEING DATA ACCORDING
TO DIFFERENT GESTALTS

Unlike the others, the response of the third investigator focused on the unusual result rather than the technical failure. In this case, the investigator wondered why cells grew better in whole blood-derived serum than in plasma-derived serum. Implicit in this question was the assumption that the observed results were correct, and the possibility was raised that the starting expectation was wrong. The investigator would want to find out if the result was repeatable; and if it were, the subsequent sequence of experiments might take place along lines entirely different from those originally planned. A new and unanticipated hypothesis for study would have emerged—namely, that some component important for cell growth is present in whole blood serum but absent from plasma-derived serum.

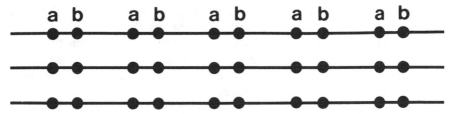

Figure 3.1   Alternative theme/background relationship I

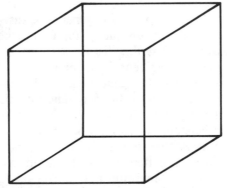

Figure 3.2   Alternative theme/background relationship II

How is it possible that investigators looking at the same data can see them so differently? The answer to this question can be resolved in part by looking at Figures 3.1 and 3.2. One can view the points in Figure 3.1 as unrelated to each other; it is also possible to see them as pairs of points on parallel lines. In the latter case, which sets of points (A-B or B-A) constitute the pairs and which the intervals between the pairs? The data can be seen either way. Similarly, it is possible to see many different images in Figure 3.2: a cube oriented up to the right or down to the left, or open at the top, and so forth. These examples demonstrate a variation in the theme/background relationship that was shown in Figures 2.2, 2.3, and 2.4. In Figures 3.1 and 3.2 the themes and backgrounds are interchangeable. Therefore, visual data presented in the field appear to mean different things depending upon which part of the field is seen as the theme.[5]

In the experimental situation under discussion, the investigators thought about the same results but each one did so according to a different (hypothetical) conceptual framework. Perhaps the first two investigators were growing cells as a source from which cell membranes were isolated to study membrane protein structure, while the third investigator was using the cells to study various aspects of growth control. In the case of the first two investigators, the fact that the cells did not grow was a nuisance. In the case of the third investigator, the fact that the cells did not grow was a clue. As has often been the case, scientific discovery depended on an investigator's ability to see data according to new gestalts that allowed previously unseen relationships among the data to become evident. This is a major aspect of the intellectual challenge of science. As discussed in Chapter 2, the data themselves can appear to be different when observed according to different gestalts.

An interesting example of seeing the same data in different contexts comes from the studies on Kuru, a degenerative disease of the central nervous system that is almost unique to the Fore tribe of New Guinea. Some scientists assumed that the disease was transmitted genetically, and two anthropologists were sent from the genetics department of the University of Adelaide to gather genealogical information that presumably would establish the genetic linkage. Their findings, however, did not clearly support the genetic hypothesis. The data were shown to an epidemiologist who was able to restate the problem in a new way: "Go out and find what it is that the adult women and children of both sexes in the Fore tribe are doing that the adult men are not doing. And whatever it is, it is something that the Gimi [an adjacent tribe] are not doing either." So the anthropologists returned to the field, and they found out that the woman and children of the Fore tribe sometimes engaged in ritual cannabilism during burial ceremonies. Thus, cannibalism was established as the mode of transmission of the disease.[6]

Alexander Fleming's discovery of penicillin is another example of how an investigator's ability to see data may be determined by the preexisting gestalts that the investigator has in mind. Unlike other bacteriologists who viewed mold as a problem and threw away their contaminated cultures, Fleming became interested precisely because the bacteria were killed by the mold. But even though Fleming used penicillin for ten years as a method for controlling bacterial growth in culture, he did not pursue clinical development of the drug.[7]

Fleming's accidental discovery of penicillin was facilitated by his previous discovery of lysozyme. At a time when he had a cold, Fleming added some of his own nasal mucus to a culture dish containing bacteria. He was looking for bacterial toxins in the mucus. He noticed a clear area in the culture dish indicating bacterial lysis, so his expectations were met. Upon further examination, however, it became evident that the lytic agent came from a contaminant in the culture and not from his own mucus. Nevertheless, since he had detected a bacteriolytic agent, Fleming pursued the observation and isolated and characterized the lysozyme. Already familiar with the appearance of a culture dish undergoing bacterial lysis, Fleming was prepared to notice a zone of bacterial lysis in a culture contaminated by mold some six years later.[8]

Fleming's lack of persistence in trying to develop penicillin into a human therapeutic agent also can be attributed in part to his studies on lysozyme. He found that lysozyme was ineffective against pathogenic bacteria, an observation that was consistent with his colleagues' ideas about the "correct" treatment of disease. Fleming was a member of the Inoculation Department of St. Mary's Hospital, London, where the primary focus of attention was on the use of human vaccination techniques rather than on drugs. Therefore, development of new drug therapies did not have a high priority.[9]

LUCK IN SCIENCE

Returning to the discussion of studies on cell growth, the third investigator might retrospectively view the use of serum derived from plasma as the lucky break that opened up a whole new approach to the field. Luck in this case meant that the investigator had the opportunity to make an unanticipated observation. Based on what previously was known and accepted, it would not have occurred to the investigator that plasma-derived serum might behave differently than whole blood–derived serum; afterward, however, the new finding seemed obvious. In this situation one often has the thought: "I don't know why I didn't do that experiment before." In contrast to the third investigator's experience, the accidental observation was not a lucky break for the others because they were not able to see the observation in the context of a new gestalt.

There is no doubt that luck is very important in science.[10] One must recognize, however, that there are different kinds of luck. One way to win a lottery is to *buy a ticket* that turns out to be a winner. Another way is to *find a ticket* that turns out to be a winner. In the first instance, the person has put himself in the winning way.[11] Sometimes, this means trying an additional experimental treatment or a new method "just to see what happens." Unexpected observations made possible by "extra tube experiments" or by application of new methodologies often prove to be the driving forces behind scientific advances. During the early development of cancer chemotherapy, the NIH was interested in screening *every new chemical compound* it was able to obtain, which is how some of the tumor-growth-regulating drugs were discovered.

When all else fails, one can always hope for a little "controlled sloppiness" in experimental execution. During the course of doing experiments, some mistakes are inevitable. The investigator may be distracted by a question and, as a consequence, use the wrong experimental buffer or interchange experimental samples. Or the investigator may become involved in an intense lunch-time meeting and forget to stop an experiment at the appropriate time. Often the investigator does not realize a mistake has been made until s/he looks at totally unexpected results. Then, if the error can be reconstructed, it may turn out to be very useful. But controlled sloppiness works only if one is always prepared to consider unusual results seriously.

One common example of luck in science is being in the right place at the right time. A biologist named Eleanor Storrs went with her husband to a conference on leprosy. He was an expert on leprosy drugs. During one of the discussions she heard that the leprosy bacillus would grow neither in culture nor in most animals, and that it liked cool places. It happened that Storrs was working with armadillos at the time, so she knew that they have a lower body temperature than most other animals. She suggested that they be tested, and it thus came about that armadillos were used as the biological factories for the growth of leprosy bacilli needed to produce a vaccine.[12]

Two other common examples of luck in science are doing the right experiment for the wrong reasons and doing the right experiment as a consequence of technical errors. Sir Peter Medawar discussed these examples in relation to his discovery of neonatal tolerance to skin transplanation.[13] He had volunteered to help a geneticist distinguish between identical and fraternal cattle twins by exchanging skin grafts between the animals. The hypothesis was that the identical twins would accept the transplants but the fraternal twins would reject them. In fact, the transplants were accepted by the fraternal twins, which seemed to make no sense.

It turned out that cattle are one of the few species in which fraternal twins have a shared blood circulation before birth, so they become exposed to each others antigens. In the context of this additional information, Medawar and his co-workers Rupert Billingham and Leslie Brent realized that they might be studying, by accident, a case of naturally occurring fetal tolerance—that is, nonresponsiveness of the immune system toward antigens to which the animals were exposed before birth. Medawar had Brent test the hypothesis directly in a controlled animal model. Fetal mice of strain A were exposed to adult tissues from mouse strain B. After birth the A mice were tested with B skin transplants. The A mice that were exposed to the B tissues before birth did not reject the B skin; they had become tolerant. The postscript to this story is that Medawar's group was inexperienced at preparing tissue fragments, and they made some fortunate technical mistakes. Had they done the experiments in a slightly different way, the host mice would have died. Billingham and Brent later made this discovery and called the phenomenon "graft-versus-host disease."

EXPLICIT AND IMPLICIT HYPOTHESES

The importance of expectations in experimental design cannot be over-emphasized. Every investigator has in mind certain hypotheses about the experimental system under study. These hypotheses will determine the investigator's expectations, and many of these hypotheses are taken for granted and never made explicit. Critical to the ability of investigators to design good experiments is their understanding that hypotheses determine not only the expected results but also the *possible* results that can be obtained. Otherwise, one is in the position of the marine biologist who, after studying deep-sea life using a net with two-inch mesh, concluded that there were no fish smaller than two inches in the sea.[14]

In general, every experiment involves selection of one or several available techniques, which are selected precisely because the investigator expects the techniques to be applicable to the system under study. In making this selection, the investigator predetermines the possible range of results of an experiment. An investigator can learn something new from the experiment only if the experimental results fall within the expected range.

In addition to determining the choice of methods, the investigator determines the control experiments to be used. There are two classes of control experiments. One class establishes the possible range of variation in the results, and includes negative controls to determine nonspecific background and positive controls to determine the level of full activity in the system under investigation. In the aforementioned experiments concerning the role of serum factors in cell growth, the method used to determine growth might be incorporation of radioactive precursors into nucleic acids. In the negative control, the investigator would omit serum entirely in order to determine how much "background" incorporation takes place. In the positive control, unfractioned serum would be added to determine the maximum amount of incorporation possible. *Establishing negative and positive controls are essential to ensuring that the system is working correctly.* If something technically goes wrong that decreases or inhibits the ability of the cells to grow, then the cells would not grow even if serum were added. If some serum accidentally was added to the cells, then the cells would grow even if additional serum was not added. Such problems with the experimental system can be detected only if one uses appropriate positive and negative controls.

It is sometimes said that there are three kinds of experiments. In the first kind, the result will be meaningful regardless of the outcome. In the second kind, one type of result will be meaningful but another will not be so. In the third kind, the results will be meaningless regardless of the outcome. The first experiment is designed so that a definitive result must be obtained. This is a situation in which the positive and negative controls are clearly defined. In the second case, either the positive or negative control is included, but for some reason not both. As a result, only one type of result will be useful. In the third case, neither positive nor negative controls are included, so no conclusions can be drawn.

The three types of experiments can be illustrated by the game of twenty questions in which the person who is thinking of something answers only "yes" or "no." The following three questions are asked: (1) Is what you are thinking of blue? (2) Is what you are thinking of blue or green? (3) What color is it?

Either a "yes" or "no" answer to the first question leads to definitive information. The second question leads to definitive information if the answer is "no," but only partial information if the answer is "yes." A "yes" or "no" answer to the third question is meaningless and leads only to confusion.

A second class of control experiments is concerned not directly with whether the experimental system is working properly but, rather, with the clarification of data interpretation. For instance, while one has the expectation that serum can be fractionated to obtain an active component, the fractionation technique may destroy the biological activity of the growth factor, or the growth factor may actually be two separate components requiring both to be present for activity to be observed. In either of these situations, the investigator would not be able to detect growth activity in the isolated fractions. A useful control, therefore, would be to recombine all of the isolated serum fractions and test the mixture. If the activity is absent from the combined fractions as well as from the individual fractions, the fractionation procedure may have destroyed the biological activity. If the individual fractions are inactive but the recombined fractions are active, then the likelihood is that several separate components are working together. Only if the investigator anticipated these possibilities would s/he have included the controls that could clarify how to interpret the results.

I must emphasize that this discussion concerns expectations as they affect the range of data that in principle can be collected in an experiment, not data collection *per se*. Students usually learn the importance of what are known as "blind" experiments. In this situation, one investigator sets up the experiment and a second investigator, who does not know the protocol, collects the data. The reason for carrying out experiments blindly is that the data collected may be influenced by an investigator's expectations. While the use of the blind procedure can correct for this tendency, it cannot change the fact that *the investigator's expectations predetermine the range of data that can be collected.*

Once the relationship between expectations and experimental design is recognized, it becomes clear why every experiment addresses both explicit and implicit hypotheses. The explicit hypothesis refers to the question the investigator hopes to answer, while implicit hypotheses determine the experimental conditions that will be used to answer the question. Thus, the explicit hypothesis under question has meaning only if the implicit hypotheses are suitable to accomplish what the investigator has in mind. Because of such potential problems with experimental design, investigators often choose to carry out experiments using more than one kind of methodological approach. The same conclusion reached by two independent approaches makes one more confident.

When the experimental design selected is inadequate to test the question posed, the consequences are confusion and delay. For instance, in the early 1900s, C.O. Jensen reported on tumor transplantation and soon many laboratories were doing mouse transplantation experiments. The results were inconsistent, and the problem turned out to be with the mice. Investigators were reporting studies on the white mouse, the brown mouse, and the spotted mouse. They had assumed, incorrectly, that mice of the same color would react in a similar fashion. The eventual solution to the problem was development of inbred mouse strains.[15]

In the studies on Kuru, one reason the disease was believed to be transmitted genetically was that no infectious agent similar to a virus could be detected. The experiments looking for viral infection used incubation times of up to one month, which would have been sufficient in most instances. An American veterinarian named William Hadlow was in the right place at the right time. He saw an exhibition about Kuru while visiting in England and recognized the possible relationship of Kuru to scrapie, a disease in which he was interested. Scrapie is a neurological disorder that effects sheep, and Hadlow knew that it was caused by a *slow-acting* infectious agent. He communicated this information to D. Carleton Gajdusek, an American physician and scientist who had spent ten months mapping Kuru throughout the Fore territory in New Guinea, and who had organized the experiments looking for an infectious agent. Under Gajdusek's direction the infectivity experiments were repeated, this time with anticipated incubation times of up to five years, and the results were positive. The role of slow viruses in Kuru was established.[16]

One other important point regarding an investigator's expectations and the choice of experimental systems is that systems may be chosen because they are the best ones available for studying a problem even if they have known limitations. In recent years, for instance, many laboratories have attempted to study what happens when normal cells become cancerous. One general approach to this problem has been to use viruses or toxic compounds to induce normal cells to become cancerous and then to measure cellular and molecular changes in the transformed cells compared to the normal cells. Most cells chosen for these studies have been fibroblasts because they are easy to obtain, easy to grow in culture, and easy to transform. Nevertheless, even though considerable new and useful information has been forthcoming from these studies, most mammalian cancers are derived from epithelial, not fibroblastic, cells. In fact, there are very few fibroblast cancers of clinical relevance. Epithelial cells, however, are harder to purify and harder to grow in culture, so they have not been studied as thoroughly as fibroblasts.

The above situation illustrates the point made earlier that development of new methods can have very significant and far-reaching consequences. The establishment of easy conditions for isolating and growing epithelial cells in culture would make it possible for many investigators to carry out experiments on the malignant transformation of epithelial cells— experiments that would have been very difficult before the new methods were developed.

DEVELOPMENT OF HYPOTHESES

In summary, experimental results can affect previously held hypotheses in three different ways.

1. The hypothesis is confirmed. (Cell growth requires serum.)
2. The hypothesis is extended. (A serum factor required for cell growth has been isolated.)
3. The hypothesis is modified or rejected. (Serum derived from whole blood contains a cell growth factor absent from plasma-derived serum.)

In the first instance, the experimental results have repeated what others had previously described. In the process, two assumptions of science have been confirmed: (1) There is a universal order underlying natural events that can be discovered by humans carrying out experiments, and, consequently, (2) this order is intersubjectively valid. "I am like other scientists and can observe the same things that they have observed." If the hypothesis had been in dispute or proposed by only one other laboratory, then the confirming experiments would lead to a more general acceptance of the hypothesis. In any event, a new series of experiments will have been linked with research carried out in the past by other investigators.

In the second instance, the factual content of the scientific domain has been extended. Novel observations have been made. Implicit in this extension is confirmation of the investigator's explicit starting hypothesis as well as his/her hidden expectations regarding the usefulness of the experimental system under study.

In the third instance, the theoretical content of the scientific domain has been extended. The modified hypothesis that the investigator has in mind represents a new way of looking at the experimental system, and may lead to a new line of experimentation that previously was unknown. Implicit in this modification is the situation that one's expectations were not met, and all or part of the starting hypothesis was called into question. The way hypotheses usually get rejected is that they are modified until nothing remains of the original idea.

Every hypothesis goes through all of the above stages at one time or another. Typically, the hypothesis is formulated in a vague way and then clarified and confirmed through additional experiments. Subsequently, the hypothesis is refined and extended. Eventually, however, observations are made that call into question part or all of the original hypothesis. As a result, it becomes necessary to conceptualize a new or modified hypothesis that can account for the new observations as well as for the original ones.

It often has been claimed that investigators develop new hypotheses by a somewhat mysterious, intuitive process.[17] For instance, the German scientist Kekulé von Stradonitz dreamed about the ring structure of the organic molecule called benzene.

> All at once, I saw one of the snakes seize hold of its own tail, and the form whirled mockingly before my eyes. As if in a flash of lightning I awoke and spent the rest of the night working out the consequences of the hypothesis.[18]

The mainstream emphasis in modern philosophy of science, therefore, has concerned the ways in which scientists validate their hypotheses rather than the ways in which hypotheses were conceptualized initially.[19]

Without new hypotheses, however, science would not advance. Consequently, an interest in the philosophy of scientific discovery persists.[20] Helpful in understanding the development of new hypotheses is the realization that the evolutionary process of hypothesis development described above is quite similar to the development of object concepts. That is, hypotheses are like schema (discussed in Chapter 2). Insofar as an investigator understands a system, s/he does so in the context of previously accepted hypotheses. New observations can be made only if they can be assimilated by these hypotheses. On the other hand, hypotheses are fluid. They can change (accommodate) in response to new data that has been assimilated. These changes are typically small, and the experimental results often lead to simple extension of a hypothesis. Sometimes, however, hypotheses are restructured as they accommodate to the experimental results.

The fluid fashion in which hypotheses function can be illustrated by the idea that scientific problem solving is something like doing a jigsaw puzzle.[21] The puzzle in this case, however, is unusual in that the outline is constrained but not fixed. That is, the outline stays the same only so long as the starting hypothesis does not change. In addition, not all of the pieces (observational data) are available. The investigator attempts to confirm the hypothesis in question by learning how to put the pieces together to form the outline. Subsequently, as new pieces are obtained, they can be added to the puzzle to fill in missing aspects. But if pieces become available that cannot be accommodated into the puzzle, then one of two things occurs. Initially, there is a conservative attempt to rearrange the pieces while preserving the outline of the puzzle. Then, if rearrangement does not work, the starting hypothesis is called into question. This leads to a change in the outline of the puzzle, and the game starts over again. Moreover, as one changes the hypothesis it may become obvious that some of the starting pieces got mixed in from other puzzles, and that some of the starting pieces were shaped differently than originally thought.

Since different data fields can potentially be seen in more than one way, it is useful to think of a current hypothesis as a gestalt that includes not only a particular arrangement of previously gathered observational data but also previously accepted hypotheses relating to the experimental system under study. The meaning of the observational data in any particular experiment is determined by the gestalt. The individual pieces of data presented in Figures 2.5A and 2.5B take on meaning according to the hypothesis used to arrange the data. Typically, experimental data seen to be meaningful at one time can be viewed later as relatively unimportant, whereas data seen initially as irrelevant can later appear to be essential. Although the "reality" giving rise to the data has not changed under these circumstances, the data will be seen differently when observed according to different gestalts.

It has been suggested that a necessary feature of scientific hypotheses is their potential for falsification. That is, hypotheses that are scientific must give rise to predictions that can be experimentally tested and potentially disproved. According to this model, science progresses through selective falsification of competing hypotheses.[22] It is assumed that no amount of confirmatory data can prove absolutely that a hypothesis is true, but also that it takes only one negative result to call a hypothesis into question. This situation is not so straightforward, however, if one considers the following three scientific hypotheses:

1. All cells capable of reproduction contain deoxyribonucleic acid (DNA).
2. Some cells capable of reproduction contain DNA.
3. Ninety percent of all cells capable of reproduction contain DNA.

The idea of falsification of competing hypotheses was proposed mainly for universal generalizations such as the first one listed above. In my experience, however, most research utilizes hypotheses like the second and third ones. In any event, the first hypothesis can in principle be falsified by a single experiment, but the second hypothesis can be confirmed by a single experiment. The third hypothesis, on the other hand, can be neither falsified nor confirmed by a single experiment.

Furthermore, confirmation or falsification presumes that suitable experiments can be carried out. This is not necessarily the case, however. As described earlier, all experiments designed to test a hypothesis presuppose other underlying hypotheses whose validity determines the adequacy of the experimental design. Therefore, falsification of a hypothesis is tentative, as is confirmation. What once was accepted as false sometimes turns out to be true. From this point of view, experiments that confirm hypotheses are just as useful and valid as those that falsify hypotheses. Indeed, most experiments are designed to extend what previously has been accepted, and the success of such experiments confirms the hypotheses under study. What makes falsification of hypotheses of particular interest is the disruption of the routine of day-to-day science that occurs. Novel results have been obtained that are puzzling, and in such cases the investigator is challenged to develop a new understanding.

It follows that *older hypotheses are not disproved so much as they are replaced by newer ones.*[23] Generally, in carrying out experiments, one uses the "best available hypothesis"—even though data may exist that indicate the hypothesis is probably incorrect in many details if

not in its overall approach. Only when a better hypothesis is conceptualized and confirmed as being more powerful in its usefulness to
understand all of the observational data is the original hypothesis
discarded. Development of a new hypothesis, therefore, has as a starting
point the acquisition of new observational data that call into question
the original hypothesis and lead the investigator to believe that the
original hypothesis was wrong even though no suitable replacement has
been conceptualized. The end-point occurs when a new hypothesis is
conceptualized and seen as explaining both the new data and the
original set of data more effectively than the starting hypothesis.

Investigators also sometimes choose between competing hypotheses
as a result of nonscientific beliefs.[24] Robert Fludd, one of the early
supporters of William Harvey's theory of blood circulation, favored
Harvey's hypothesis because it supported Fludd's belief in the basic
parallel between the cosmos (circular motion of the planets) and man
(circular motion of blood).[25] By the same token, the Copernican notion
of the sun as the center of the universe (Chapter 1) was readily accepted
by Kepler because it agreed with his religious ideas.

> In the first place . . . of all the bodies in the universe the most
> excellent is the sun, whose whole essence is nothing else but light, than
> which there is no greater star . . . most fair, limpid, and pure to the
> sight . . . and which alone we should judge worthy of the Most High
> God, should he be pleased with a material domicile and choose a place
> to dwell with the blessed angels.[26]

STRUCTURE, FUNCTION,
AND ORGANIZATION IN BIOLOGY

Biologists develop hypotheses regarding both the structure and the
function of biological organisms. The meaning of function in this context
needs to be clarified. Consider the following three statements about
mammalian hearts:

1. The purpose of the heart is to pump blood.
2. The function of the heart is to pump blood.
3. An effect of the heart is to pump blood.[27]

The first and second statements are sometimes dismissed because
of their teleological implications. In the first statement, for instance,
the language can be mistaken for an analogy with conscious, purposive
human behavior. While the purpose of artificial hearts might be to
pump blood (the purpose belonging to the artificial heart's inventor or

the surgeon who implants it), hearts as they have developed evolutionarily have no conscious purpose insofar as one is aware. The third statement is a causal form in which nothing more is expressed than an observation of what hearts have been observed to do. From this statement, however, one cannot understand the signficance of that particular activity of hearts with respect to an organism as a whole. This is the point of the second statement. That is, hearts not only pump blood; they also have a special relationship to mammalian organisms. This relationship is precisely what is implied when one states the heart's function.

To emphasize the above point, consider the possibilities for describing the sounds that hearts make. Although one would be correct to say that an effect of the heart is to make heart sounds, it would make no biological sense to say that the function of the heart is to make heart sounds. Because biological organisms are interacting, multicomponent systems, it is necessary to use functional language in describing them.

It is also important to understand that interacting systems can be analyzed at different organizational levels. For instance, the specific aims of the laboratory interested in studying the growth control of mammalian cells were to isolate growth factors, to determine their mechanism of action at the cellular level, and to determine normal physiological function. These aims encompass three different levels of organization: molecular, cellular, and organismic.

The mammalian organism is composed of many different cell types organized into specific tissue and organ systems. To understand cell function, one must remove the cells from the normal situation. Having done so, one can learn about how cells function *as individuals,* but not precisely which cell functions are expressed *in the organism.* The latter depends on how cells of the same type are organized together, and how these cells are influenced by cells of other types. Investigators often choose to study problems at either the molecular, cellular, or organismic level because it is very difficult for one laboratory to study a problem from all three points of view. Nevertheless, the information developed from any one perspective is incomplete. Studies on individual cells can potentially determine all of the different factors that can influence cell growth, but only when the arrangement of cells in the organism is taken into consideration will one begin to see which of the factors are operational and under what circumstances.

Similarly, individual cells are multicomponent systems composed of organelles and cytoplasm elements interacting with each other and with the external cellular environment. (See the discussion of cells as compartments in Chapter 2.) The study of the activities of these organelles or cytoplasmic elements often requires their isolation from their normal environment (i.e., the remainder of the cell). Again, however, only the range of potential activities can be determined by studies on the isolated organelles. At the other end of the biological scale, in evolutionary biology, the problem of organization is also complex.[28] The individual organisms of a species establish its genetic composition. But when it comes to natural selection, the geographical distribution of organisms may be as important as their genetics. That is because environmental disturbances often act locally, not globally.

A good example to illustrate the importance of organization in biology is the relationship between protein structure and function. Proteins are part of all biological organisms, down to very small viruses. Each protein molecule is a polymer composed of about twenty different types of amino acids arranged in a particular sequence, which is called the primary structure. In addition, the polymer can undergo molecular folding and interchain and intrachain cross-linking, all of which lead to additional degrees of structural organization. The crucial point is that the activity of the protein is determined by the amino acids present, their sequence in the polymer, and the folding of the polymer. Determining the quantity of each amino acid in a protein is only of limited usefulness in predicting the protein's activity or structure. The situation is analogous to the relationships among letters, words, sentences, and paragraphs. Imagine being presented with the letters "a,t,c,s" and asked to predict the possible words that might be spelled. There are twenty-four possible combinations, four of which are quickly recognized as having meaning. While the letters available set the boundary conditions for which words can be spelled, the arrangements of the letters determine the meanings of the words. For a simple protein, which might have fifty amino acids in the polymer, an enormous number of sequences are possible, even without taking into consideration the higher order structure.

A useful way of thinking about this situation is to realize that the components of a biological system determine the biological boundary conditions of the system, while the arrangement of the components determine where within the boundary conditions particular biological activities are expressed.[29] Many scientists claim to believe in a reductionist approach to understanding nature—that is, the idea that events can be understood and explained by studying them at more and more fundamental levels. In practice, however, biologists investigate not only the components that make up systems but also the arrangements of the components in the intact system (the latter is an example of the holistic approach). After isolating and studying the components of a biological system as individuals, one tries to put them back in proper relationship to each other in order to understand how the system operates as a whole.

Scientific Collectives: Transmission of the Thought Style

4 Science begins with the individual investigator, who makes observations and designs and interprets experiments. Then, the investigator brings the results to the attention of his/her scientific peer group. The members of the peer group are responsible for carefully and critically evaluating the work. Therefore, it is important to understand science as a social activity in which individuals are engaged because, in the final analysis, what makes the observations and experiments of one investigator scientific is their acceptance by others. And such acceptance is not trivial because scientists take seriously the motto of the Royal Society, *Nullius in verba,* which P. B. Medawar translates: "Don't take anybody's word for it."[1]

This chapter and the next are about the peer relationships among scientists. These relationships will be described in terms of scientific thought collectives and thought styles. After explaining the structure of thought collectives, the discussion in Chapter 4 will focus on the ways in which they transmit their thought styles to new members, particularly students. Then, in Chapter 5, the function of collectives to set scientific standards will be described.

THOUGHT COLLECTIVES
AND INTERSUBJECTIVITY

According to the scientific attitude, the observations and hypotheses of science should be universally valid because it is assumed that there is a universal order underlying natural events. Universality, however, cannot be determined by a single investigator, so the observations and hypotheses proposed by an individual investigator must be confirmed or confirmable by others. *Although individual investigators carry out scientific activities, scientists as a group either accept or reject the activities as being scientific.*

My assumption as an individual scientist is that other investigators could adopt a similar scientific attitude and confirm my observations and hypotheses. Thus, I take it for granted that my work is intersubjectively valid. This belief is a natural outcome of my typical experience of the world. I experience the world as present not only to me but to others as well. Moreover, I assume that others' typical experience of the world is similar to mine. For instance, during class my students and I can agree that we are in the same room at the same time, and that there is a single large table around which we all are sitting. I assume that any variations in what we see are a consequence of differences in our past experiences or in the perspectives from which we are looking. Typically, everyone has this experience of sharing the same situations with others. It is my experience of the world as being *ours, not mine alone* that makes it natural for me to anticipate that I share with other investigators a scientific view of the world.

In order for investigators to confirm that the scientific domain is shared, it is necessary for them to communicate with each other. When an investigator tries to explain scientific ideas to another investigator, a medical student, or an elementary school student, the level of communication needs to be appropriate to the audience. Unless the speaker converses in terms that will be suitable to the background of the listener, the presentation will be ineffective. Similarly, in all interactions, individuals communicate with each other according to typical expectations of what sort of conversation would be appropriate to a particular situation. Groups of individuals who share common interests and activities (social, political, religious, scientific, and so forth) also tend to share typical expectations and a common language used in carrying out group activities.

A group of individuals engaged in some activity may have exclusive beliefs about the meaning and proper performance of their activity, and these beliefs may be regarded as absolute truths. Such exclusivity is often a characteristic of religious and political groups. Individuals who are members of one such group may tolerate participants in other, opposing groups while viewing the members of the other groups as being misled. Alternatively, one group may attempt to destroy the beliefs of another by physically eradicating or subjugating the members of the group. Scientific groups are unique in that the ideal goal of science is inclusive knowledge. That is, truth in science aims toward consensus, and it is based on the idea that everyone potentially could verify this truth. *The scientific attitude, unlike other attitudes toward experience, believes itself to be radically intersubjective.* Simply put, most scientists do things in ways that they assume will be believable to other scientists. Consequently, even the most basic features of what counts for scientific evidence change depending upon what kinds of evidence investigators think will be convincing to each other.[2]

Scientific groups are variable in size and function, and an individual investigator can be a member of many different groups. At the local level, for instance, I am a member of the following series of groups: my laboratory, my academic department, and my university. In a broader sense, I am also a member of a series of groups that might be designated: my laboratory, all investigators studying problems similar to those that I study, and all investigators in the field of cell biology. In the local context, the primary function of the groups is to recruit and educate students. That this is the case can be demonstrated by what happens to a university and its faculty members when student enrollment declines. When too few students sign up for a program, the program is dropped. In the broader context, the function of the groups is to carry out scientific research.

Every group is characterized by certain accepted attitudes regarding the way in which the activities of the groups should be carried out— for example, how students are to be educated or how scientific problems should be defined and solved. These attitudes constitute the operating "paradigm" of the group.[3] Another way of expressing a similar idea is that such groups comprise "thought collectives" that share a common "thought style" about how group activities should be carried out.[4] I often will refer to scientific thought collectives in the singular, but the reader should keep in mind that there are as many different thought collectives as there are different organizational levels within science.

Referring to a scientific group as a thought collective is not meant to define a social structure. Rather, the purpose is to focus attention on those assumptions held in common by individual investigators who are interacting with each other for some reason. Each investigator has a particular background knowledge that is unique, but there are some aspects of the background knowledge that are shared with other investigators. This shared knowledge includes assumptions about various aspects of science including methodological approaches, observations, accepted hypotheses, and important problems requiring further investigation. These shared assumptions and beliefs are the prevailing thought style of the collective, and they include a definition of what it means to do research. As new observations and hypotheses are assimilated by the thought style, it changes to accommodate them. *To be acceptable, the new material must appear to be a natural extension of the old.*

Novel observations and hypotheses that are discontinuous with the thought style are usually discounted, at least temporarily.

One interesting feature of scientific thought collectives is their anonymity. Most investigators do not think of themselves as participants in a thought collective. This anonymity vanishes, however, whenever conflicts arise. Then the battle lines (fortunately intellectual) are drawn between different "schools of thought." In addition, the scientific collective as a whole can come into confrontation with thought collectives outside of science. In the 1983 Arkansas court case concerning how the origins of the earth and life on earth should be taught in the public schools, much of the testimony centered on the question of whether Creation Science, which is favored by the religious right (i.e., the fundamentalist thought collective), is scientific.[5]

Members of the fundamentalist thought collective say that they utilize only "scientific data," but they use the word *science* in a way different from that used by the thought collective made up of biologists. Creationists take for granted that there only are two possible hypotheses regarding the origin of species, evolution and creationism, one of which must be correct. Based on this assumption, they conclude that any evidence not supporting evolution favors creationism. Unlike the theory of evolution, which is based on observations that are intersubjectively verifiable, Creation Science is subject neither to observation nor to intersubjective verification.[6] Thus, these two different views of "how things really are" involve different thought styles and different assumptions about what doing "science" means.

The views held by members of scientific collectives change over time according to the prevailing thought style. In 1966, Peyton Rous won the Nobel Prize for his work demonstrating the viral transmission of cancer, research that was carried out in the early 1900s. George Klein, in presenting the award to Rous, noted that other investigators had failed to obtain similar results with mice and rats, and concluded that chicken tumors were curious exceptions to the normal mode of tumor causation. It was not until the 1950s, when other scientists discovered that the polyoma virus could cause tumors in many different mammalian species, that Rous's findings took on new significance. After almost fifty years as a curiosity, the idea that tumors were virally transmitted became central to the thought style of cancer researchers in the 1960s.[7]

It was not long afterward, however, that the thought style began to change again. The discovery of viral particles in the cells of most vertebrate species led to the idea that the viruses were transmitted from parent to offspring, and that tumors occurred because the "oncogene" portion of the endogenous virus was activated somehow. Then, it turned out that uninfected vertebrate cells contain cellular genes that are very similar to the viral oncogenes, and the conclusion was drawn that many tumors may occur as a result of altered regulation of normal cellular genes.[8] Thus, the idea that most tumors are transmitted by viruses—briefly a leading hypothesis in understanding malignancy—became less likely.

ACCEPTANCE OF NEW DISCOVERIES
BY THE THOUGHT COLLECTIVE

In recent years, there has been an increasing interest in understanding the resistance of scientists to new discoveries. The development of modern enzymology provides good examples of resistance and shifts in attitudes in science.[9] The belief in a distinction between living and nonliving matter led to the assumption that organic compounds could not be synthesized in the same way as inorganic compounds were synthesized. Even after urea was synthesized and the theoretical distinction between inorganic and organic compounds was disproved, a lingering belief in vitalistic forces had to be overcome before enzymes could be accepted as the catalysts responsible for the complex reactions typical of biological organisms. Once this controversy was finally settled through reproduction of the process of fermentation *in vitro* with a cell-free yeast extract, there was still some reluctance to accept enzymes as examples of already known organic compounds (e.g., lipids, carbohydrates, and proteins). Finally, this point was settled by isolating, crystallizing, and characterizing the enzyme pepsin.

Max Planck is supposed to have commented that scientists never change their minds, but eventually they die. Thus, it may require a new generation of scientists to accept what the previous generation resisted. Darwin implied as much in "On the Origin of Species" when he wrote:

> A few naturalists, endowed with much flexibility of mind, and who have already begun to doubt the immutability of species, may be influenced by this volume: but I look with confidence to the future, to young and rising naturalists, who will be able to view both sides of the question with impartiality.[10]

An analysis of the positions of British scientists vis-à-vis Darwinism has shown that ten years after the theory of evolution was proposed, it was accepted by scientists of all ages. Age factors could account for only a small percentage of those scientists who did not accept the theory. In the above study, however, an investigator joined the class of those accepting evolution if he stated explicitly that he believed in evolution (i.e., the evolution of one species into another), regardless of his beliefs about the mechanism.[11]

Whether or not scientists accept a new hypothesis depends in part on what one means by *accept*.[12] In a 1976 review of his "chemiosmotic theory" of oxidative phosphorylation, Peter Mitchell published a chart showing the attitudes of other investigators toward his theory between 1961 and 1973.[13] A comprehensive sociological study on the development of ideas about oxidative phosphorylation included interviews with protagonists in the field, some of whom were asked to comment on Mitchell's chart.[14] Several individuals agreed in a general way with Mitchell. Others said that they were fully convinced that Mitchell was half-right, or that Mitchell had published different versions of his theory and that it would be possible to state whether or not one agreed only if a particular version of the theory was specified.

Writing about his chart, Mitchell said:

> It is understandable, however, that some of the eminent biochemists, who have long championed the more traditional . . . views of the coupling mechanism, have not found it easy or agreeable to acquire a taste for . . . disciplines . . . that were not originally of their own choosing.[15]

The important point is that scientists who have been involved in a research problem from its early development *know too much* to be able to accept a new hypothesis easily and completely. As a consequence of their experiences, they are familiar with many nuances of the evidence about which younger investigators are naive. Moreover, the older scientists tend to have a vested interest in the dispute.

The textbook of biochemistry that I used in 1966 discussed only one mechanism of oxidative phosphorylation—namely, chemical coupling.[16] A student using A. L. Lehninger's textbook in 1975[17] could have read in detail about three different possible mechanisms of oxidative phosphorylation, and learned that although the other mechanisms could not be ruled out, the chemiosmotic hypothesis accounted most simply and directly for the experimental evidence. The 1982 version of the same textbook taught that the chemiosmotic hypothesis was "widely accepted," although important problems remained to be resolved. Thus, unlike students educated in the early 1960s, those educated in 1982 never experienced the chemiosmotic theory as a controversial hypothesis. If one wants to follow the development of consensus in science regarding a controversial issue, the place to look is in the textbooks covering the field.[18]

The resistance of scientists to new ideas sometimes comes from beliefs outside of science. For many people the idea that the world was round did not fit with common-sense experience. The earth does not look round and does not seem to be moving. Also, ideas that at first were explained within the nonscientific domain of experience may not easily be incorporated into the scientific domain. Syphilis was understood initially to be an astrological disorder or a religious punishment. Both of these views had to be overcome before syphilis could be recognized as an infectious disease.[19]

From within science, hypotheses usually are discounted for one of three reasons. First, the new ideas may be in conflict with other ideas that previously were accepted. Investigators willing to accept William Harvey's theory of blood circulation put forth in the 1600s had to reject Galen's theory of blood synthesis that had been accepted for the previous 1,400 years.[20] Second, new ideas may be opposed because they utilize new methodological approaches. Harvey measured the amount of blood passing through the heart and concluded that it could not have been synthesized by the liver quickly enough. These measurements were the first analytical experiments carried out in physiology, and Harvey was criticized for acting like a logistician instead of an anatomist.

When microscopes were introduced into the medical school anatomy course at Oxford University in 1845, Dr. W. Tuckwell described the reaction of the senior faculty as follows:

[The lectures] were delivered in the downstairs theatre, whence we ascended to the room above, to sit at tables furnished with little railroads on which ran microscopes charged with illustrations of the lecture, alternatively with trays of coffee. A few senior men came from time to time, but could not force their minds into the new groove. Dr. Ogle, applying his eye to the microscope, screwed a quarter inch right through the object, and Dr. Kidd, after examining some delicate morphological preparation—made answer first, that he did not believe in it, and, secondly that if it were true he did not think God meant us to know it.[21]

A third scientific reason why new ideas are resisted is that they do not fit in with other, previously accepted hypotheses. The lack of understanding of Peyton Rous's research was one example. Another is the work of Barbara McClintock. Her studies on transposable genetic elements were published in detail in the 1950s. McClintock's findings, though believed, generally were not viewed to be of major consequence for understanding the "typical" mechanisms of genetic variation. Once the thought collective of molecular biologists had advanced to the point that it was able to understand the significance of jumping genes, the central importance of McClintock's work was recognized. In 1983 she was awarded the Nobel Prize in Physiology or Medicine, which most biomedical scientists would agree is the highest honor that a scientist could receive.

From the vantage point of the historical observer, it is only when a scientific thought collective is forced to undergo major changes in thought style that it becomes easy to recognize the role of the collective in setting the normative assumptions of science. Such periods in the history of science when major changes occur have been called "scientific revolutions," in contrast to periods of "normal science."[22] The difference is quantitative. Revolutions occur routinely during the course of every-day science, and slight changes in the thought style are taking place continuously.

STRUCTURE OF GRADUATE PROGRAMS

Continuity of the scientific collective requires transmission of the thought style from old members of the collective to new recruits. Typically, students are exposed to the views of thought collectives throughout their academic training. From elementary school through college they become acquainted with the basic scientific knowledge that is part of the broad thought collective that might be called science-in-general. During their first two years of graduate school, they come under the influence of graduate programs that represent much more specific disciplines. Such graduate programs are thought collectives made up of faculty members whose research interests lie within the scope of the discipline. After demonstrating mastery of the graduate program thought style, students work full-time in individual laboratories where their thesis research is carried out under the supervision of one of the faculty members of the graduate program.

Graduate programs usually are composed of both senior and junior faculty members. The senior faculty members, through their previous participation, represent the prevailing thought style of the program. The junior members, who have joined the program more recently, usually occupy a more marginal status with respect to the collective's thought style. As carriers of other thought styles from their previous experiences, the junior members may view the thought style of the established graduate program as being in need of modification. Thus, social collectives and thought collectives are not necessarily identical. Faculty members of a graduate program constitute a formal social collective, but those young investigators who have recently joined the social collective must adopt or integrate their own thought styles into the prevailing thought style before they become part of the thought collective.

The organization of scientific collectives is typical of most collectives experienced in everyday life. There is a core, inner group that has the greatest power to set policy (i.e., implement the thought style). Surrounding this core is an outer group of individuals who are in various stages of movement toward the core based on their demonstrated practice of the thought style. In collectives where the core group has the power to eject unwanted individuals, dismissal usually occurs when a person fails to grasp the prevailing thought style. Senior faculty members, for instance, decide whether a junior faculty member can continue as part of an academic department based on whether the junior member is viewed as training students and carrying out research in a manner consistent with the accepted thought style.

When several competing thought collectives participate in the same graduate program, potential conflict may develop regarding the means by which various aspects of the program are to be carried out. Discussions held in an attempt to resolve such conflicts are interesting in that they are rarely scientific. That is, individual investigators infrequently (never in my experience) point to or look for sociological studies indicating the potential success or failure of different methods of graduate education. Rather, the arguments made are typically statements of past experience ("this is a method that I'm familiar with that has worked before")—statements that are really appeals for continuing a particular thought style.

Occasionally, some aspects of graduate education are analyzed in depth, such as in the 1984 report on medical school curricula by the Association of American Medical Colleges.[23] Central to the conclusions of this report is the implication that medical students spend too much time learning factual knowledge and not enough time learning skills, values, and attitudes. This report potentially could lead to major changes in medical education, but thought collectives resist change. The same concerns put forth in the 1984 study were raised by similar studies on previous occasions without much impact.[24]

THE REPUTATION OF GRADUATE PROGRAMS

Every graduate program sets admission requirements that are standards according to which applicants to the program will be evaluated. These standards are partly quantitative (grade-point average and Graduate Record Examination scores) and partly qualitative (local interviews, letters of recommendation, and prior research experience). In some programs, the faculty members emphasize the students' academic achievements ("brighter students make better scientists"). Other programs, however, are more concerned with a student's prior experience ("this student has worked in a laboratory and is still interested in pursuing a career in science, which indicates his/her motivation"). Inexperienced students who are admitted into biomedical graduate programs may find out later that they do not enjoy doing research and drop out to enter medical school or other activities.

The actual standards used for admitting students to the graduate program are determined by two major criteria. On the one hand, it is desirable to recruit the very best applicants. These are the students whose accomplishments are believed to predict most closely their future success as researchers. On the other hand, the program recognizes the limitations of the available applicant pool. If the admission standards are set too high, then few or no students will join the program. Since survival of the thought collective requires a continuous supply of new members, a compromise has to be reached between what is desirable and what is possible, taking into consideration the collective's "self-image."

If the graduate program faculty members are young, newly established investigators, they are in a marginal position with respect to the thought collectives of their respective research fields. With time, however, they may become well-established investigators. This transition becomes evident when they are invited to give seminars at other schools, present formal talks at meetings of research societies, serve on editorial boards of journals, and join scientific advisory panels involved in reviewing grant applications. As the scientific reputation of the collective increases, the self-recognition of this change will lead to the expectation that better students will wish to be trained in the program.

Some students apply to a particular graduate program for reasons unrelated to science (e.g., the university is in the city where my family lives, or where my husband or wife has been transferred). Others learn about graduate programs from college advisors or individuals who the students believe are familiar with graduate education. The features used to evaluate graduate programs are diverse. For instance, the National Academy of Sciences (NAS)[25] uses the following criteria:

1. Size: number of faculty, number of students, number of graduating students
2. Student characteristics: grant support, time required for training, postgraduate commitments
3. National reputation: faculty, effectiveness of educational program, recent improvements
4. University library size

TABLE 4.1
Doctoral Training of NAS Members, Nobel Laureates, and Ph.Ds in Science

University	% NAS Members (1900–1967)	% Laureates (1901–1972)	% Science Ph.Ds
Harvard	17.6	16.2	3.6
Columbia	9.2	14.9	4.3
U. C. Berkeley	6.6	8.1	4.5
Johns Hopkins	9.6	8.1	3.8
Princeton	3.8	8.1	1.9
Total	46.8	55.4	18.1
21 universities	87.8	100.0	68.1

Adapted from: Zuckerman, H., 1977, "Scientific Elite: Nobel Laureates in the United States." Macmillan Publishing Co., New York.

5. Research support: percentage of faculty with grants, total dollar level
6. Publication record: number of publications, impact of publications

Of these criteria, it is the national reputation that is most visible. That is what one typically finds in articles with titles such as "Best Graduate Programs in the United States."[26]

One way that graduate programs and universities can attain high visibility and reputation is to have their graduates become part of the scientific elite. There were approximately 200,000 Ph.D. scientists in the United States in 1976, of whom about 1,200 (0.6%) were members of the NAS—the most prestigious scientific society in the United States—and 76 (0.04%) were Nobel Laureates. The universities at which most of the NAS members and Laureates did their graduate work are listed in Table 4.1. It can be seen that 18% of the NAS members and 16% of the Laureates received their graduate training at Harvard; but, overall, Harvard awarded fewer than 4% of the doctoral degrees in science. Five universities together produced half of the NAS members and half of the Laureates but only 20% of the science Ph.D.s, and 21 universities accounted for almost all of the NAS members and Laureates.

While graduate programs with prominent reputations do not necessarily provide better graduate training than other graduate programs, students who get into and graduate from the highly visible programs improve their pedigrees through association. In addition, such students have an increased likelihood of working with an advisor who will become, or already is, part of the scientific elite. Regarding the Nobel Prizes awarded between 1901 and 1972, about 14% of the award-winning research was carried out at Harvard, 10% at Columbia, and 9% at the University of California at Berkeley (U. C. Berkeley). Moreover, five years after they were awarded the Nobel Prize, 13% of the Laureates were on the faculty at Harvard, 13% at U. C. Berkeley, and 7% at Columbia.

GRADUATE COURSES AND EXAMS

Most graduate students begin their graduate training by taking one or several general survey courses. The material covered in these courses usually encompasses the entire field under study, even if the research interests of faculty members in the graduate program do not encompass the field. The goal of the graduate program is to transmit to students the observations, hypotheses, and problems that are currently accepted by the thought collective of the field-in-general.

Students also usually take advanced courses. These tend to be selectively oriented toward the particular research interests of the program faculty. Ideally, the faculty of graduate programs would have research interests encompassing the entire field, but often this is impossible because of practical limitations in faculty salaries and space. As a consequence, a graduate program may be weak in some research areas and strong in others. The goal of advanced courses, therefore, can be viewed as an attempt on the part of the local thought collective (the graduate program) to transmit its particular interests to the students. Since the advanced courses are taught in more detail than the survey courses, the students become more familiar with the local thought style than with the thought style of the field-in-general.

After completing their formal coursework, graduate students usually are required to take qualifying examinations. Passing the qualifying examination permits the student to "advance to candidacy." Prior to this time, the students are not yet Ph.D. candidates in the eyes of the graduate program. Rather, they are in a probationary period during which their suitability to continue in the graduate program is determined. Generally, qualifying exams are comprehensive in that they are designed to test a broad range of knowledge. After passing the exams, students spend almost all of their time in individual laboratories carrying out the research that will be described in their dissertations. Except for the dissertation defense, the graduate program gives up most of its control over the students' education. By passing the qualifying exams, students demonstrate that they have satisfactorily learned the thought style of the graduate program collective.

It is noteworthy that the format of the qualifying examinations can vary considerably depending upon how precisely the graduate program collective wants to be assured that the thought style has been mastered. Implicit in the format used is the view of the collective toward its individual members. Every individual laboratory operates according to a particular thought style of its own. The thought style of individual laboratories may overlap with the graduate program thought style, but they may also differ in significant respects. The more rigorous the qualifying exams, the less the graduate program has to depend on individual laboratories to transmit the graduate program thought style to the student.

In some graduate programs, one component of qualifying exams is the thesis proposal in which a student conceptualizes and submits for approval to the qualifying exam committee a general description of the research project that s/he plans to carry out. This proposal includes a description of why the project is worth doing and what methodological approaches will be used in the proposed studies. The investigator who will be the student's advisor usually has significant input into this proposal because it is in the advisor's laboratory that the work will be carried out. What can develop, therefore, is a situation in which the evaluation of the thesis proposal is actually an evaluation of the advisor's thought style from the point of view of graduate program thought style. This creates the possibility for direct confrontation between the two thought styles, with the student caught in between.

Many students believe that the thesis proposal is a commitment to actually accomplish the work proposed. It often comes as a surprise to them that, as in the case with every grant proposal, the content of the thesis proposal serves only as a point of departure for the research. Typically, the project will evolve in ways that previously were unanticipated. The research problem solved may not be the same as the problem originally proposed for solution. The question for the qualifying examination committee is not really whether the specific research project proposed is acceptable but, rather, how far the student has progressed in being able to understand the reasons for selecting a particular research project and in being able to conceptualize methodological approaches suitable for performing the studies. Moreover, the committee seeks to learn the extent to which the student is able to articulate these ideas in a written format and to defend them orally. In part, therefore, *the student's ability to communicate with the collective is under examination.*

The same objectives also could be accomplished if the student prepared a grant proposal on a subject other than the planned thesis research. In this case, the student could generate the proposal completely on his or her own, and the student's advisor could play the role of member of the qualifying examination committee rather than that of one who is examined by the committee. Nevertheless, from the student's point of view, the closer the grant proposal is to the planned area of actual research, the more useful it will be in getting the thesis research under way.

There is a considerable difference between the way the thought style is transmitted in graduate school and the previous education of most students. College courses generally are lectures in which the course content is closely controlled. Even with seminar courses, the goal of the instructor usually is to convey to the students a certain set of known information. In graduate school, however, the content of many courses is variable. In one typical format, the course is organized around discussions of original research papers in which faculty members lead the discussions but allow the students to try to figure out the important points. In another format, faculty members simply assign topics on which students must present seminars to other students.

Both of the above types of courses teach skills that students will need later in their careers. Investigators usually learn from research seminars or from reading the literature and synthesizing the current state of a subject based upon their reading. Inviting students to think as independently as possible and to reach their own conclusions prepares them for what they will need to do in their future careers. A similar situation occurs when students are asked to prepare research review papers or mock grant applications, both of which are typical of the kinds of writing that the students will need to do later on in order to succeed as investigators.

The above course and exam formats present the student with a situation in which there usually is more than one possible right answer. Prior to graduate school, students know that they are judged based on their ability to learn factual information that can be found in textbooks. In graduate school, textbook knowledge is necessary, but it is only a beginning. Viewed from outside science, the search for new knowledge has a mystique that makes that search appear very attractive. Once engaged, however, many students find themselves committed to an intellectual challenge with no sure-to-work methods or guarantees of success. As a consequence, many students go through a depressing period—often in the second year of graduate school—that has been called the "sophomore slump." The fact that students may be probationary during this time increases the sense of uncertainty. Eventually, students either accept the uncertainty of science or drop out of graduate school.

An important goal of graduate education is to teach students how to communicate with scientific thought collectives since it is the collective that determines the scientific merit of an individual's work. If an investigator fails to communicate properly with the collective, then the work will not be viewed as scientific even if it is otherwise appropriate to the thought style. There are three typical ways by which investigators communicate with the thought collective (each of which must be carried out according to standards set by the collective): research talks, published research papers, and research grant proposals. In addition, just as communication involves transfer of information from the investigator to the collective, it also involves transfer of information from the collective to individual investigators. Therefore, students must learn how to listen to research seminars, how to ask questions, how to read research papers, and how to evaluate research grant proposals.

By giving students the opportunity to present seminars and write papers modeled along the lines of research publications and grant proposals, the graduate program is providing the students with the intellectual tools needed to communicate with the thought collective. By giving students the opportunity to hear seminars and read and critically study original research literature, the graduate program is training students how to listen to the communications of the collective.

THESIS ADVISORS

A major transition occurs in a student's situation when s/he leaves the supervision of the overall graduate program and enters the smaller thought collective of an individual research laboratory. Because a student's research advisor will have so much influence on the way in which the student learns to do science, the choice that the student makes often will be critical to his/her future. The process of choosing a thesis advisor begins at the time students enter a graduate program, or even earlier. Sometimes a student chooses a particular graduate program because s/he wants to work with an investigator with whom the student became familiar while still in college. Students get to know the faculty through informal interactions and in courses taught by the faculty members. In addition, students may spend time working in different laboratories where they are introduced to the particular problems under study and the relevant methodological approaches being used.

Some students select a particular laboratory to work in because of the research area under study. Other students choose a particular advisor because of the advisor's scientific reputation. Students recognize that in the future they will be looked upon by the thought collective, at least in part, in terms of their scientific pedigree, which will be enhanced if their advisor is well recognized. Finally, some students choose an advisor simply because they like the person as an individual.

In choosing an advisor, students often forget to ask certain questions and end up disappointed with their subsequent experience. For instance, I knew one student who hoped to finish his doctoral work in four years, but went to work for an investigator who had the reputation of keeping his students for at least six years—and, in fact, the student spent seven years getting his degree! In my role as first-year advisor, I tell my graduate students to think about the six questions listed below before choosing a research advisor. The relative importance of these questions depends on the student.

1. Is the thesis problem interesting?
2. Is the thesis problem well defined, and how long should it take to accomplish?
3. Can I get along with the advisor and his/her way of dealing with students?
4. Is the advisor interested in me?
5. Is this laboratory the place where I want to spend most of my time during the next four or more years?
6. Will the advisor be able to help me find a position in the future?

In choosing an advisor, many students overlook the special relationship between teacher and student that will develop. There is a story about a mystical religious teacher called Leib, son of Sarah, who said that he went to study with his great teacher, the Maggid, not to learn details but to see how the Maggid tied his shoelaces.[27] The relationship between student and teacher in science often has a similar quality. One Nobel Laureate described his experience with his teacher (also a Laureate) in this way:

> It's the contact: seeing how they operate, how they think, how they go about things. (Question: Not the specific knowledge?) No, not at all. It's learning a style of thinking, I guess.[28]

In summary, regardless of why advisors are selected, it is in the advisor's laboratory that the student will learn the strategy for doing research—that is, how to approach experimental design, implementation, and interpretation. Since graduate students rarely study philosophy of science as a part of their education,[29] it will be from their advisors that they will learn the goals of science and the art of distinguishing scientific experiments and hypotheses from nonscientific ones. They will learn these lessons at a practical rather than at a theoretical level by imitating the examples set by their advisors.

STRUCTURE OF
INDIVIDUAL LABORATORIES

Academic research laboratories usually consist of a senior investigator and his/her assistants. The number of assistants can vary dramatically depending on several factors, particularly the amount of space and funding available. An investigator who is held in high regard by his/her institution often is allocated more laboratory space than another who is viewed less favorably. Similarly, an investigator whose work is highly regarded in the research field is likely to obtain more funding than another whose work is appreciated less.

Available space and funding certainly influence the number of people working in the laboratory, but other factors are important as well. As the laboratory grows, the ability of the senior investigator to interact with the staff diminishes. In larger laboratories, the senior investigator may have to spend more time on administrative activities necessary to keep the laboratory running and less time on doing experiments personally. Depending on the inclinations of the investigator, therefore, a laboratory may be kept small even if the possibility exists for it to grow larger.

The members of research laboratories other than the senior investigator typically fall into the categories of technical assistants, students, and postdoctoral fellows. Technical assistants usually are closely supervised by a more senior member of the laboratory. Often, they do not play a role in the selection of research problems or participate in communication with thought collectives outside the laboratory. Few technicians are asked to present seminars or research papers at scientific meetings, and in most cases they are not listed as co-authors on research publications. Their contributions to the research tend to be acknowledged in the back sections of the papers. An important point is that many technical assistants have the ability to make the same kinds of contributions to a laboratory as graduate students or postdoctoral fellows. The difference between technical assistants and graduate students or postdoctoral fellows lies in the job description and the typical expectations that are based on these descriptions. Technical assistants are not viewed as aspiring to careers as independent researchers.

When graduate students first enter a laboratory, they tend to be in a position similar to that of a technical assistant. That is, they are told what sort of experiments to carry out. As time goes on, however, their goal will be to master the other aspects of doing research, which include designing and interpreting experiments, choosing research problems, and communicating with the thought collective. In short, graduate students enter laboratories as technical assistants and leave as post-doctoral fellows.

Postdoctoral fellows are investigators who have recently finished their doctoral degrees. The high frequency at which Ph.D.s are taking post-doctoral positions is a relatively recent phenomenon. In 1960, 20% or fewer of the graduates in the biological sciences, chemistry, and physics planned on taking postdoctoral positions. By 1978, however, that number had grown to 50% or more. This change occurred, in large part, because there were more science Ph.D.s than traditional positions available. The National Science Foundation predicted that the 1987 shortfall in life sciences jobs would be 12,000, representing about 11% of the total number of life scientists.[30]

Obtaining a faculty position depends on how a young investigator is viewed by the thought collective, which in turn depends primarily on research accomplishments and the ability to communicate these accomplishments. Interviews conducted in several major U.S. universities revealed that in 1978, 73% of the applicants for assistant professorships in the life sciences had postdoctoral experience.[31] The postdoctoral position has become an *expected* period of additional training and exposure during which the young investigator has an opportunity to increase and demonstrate mastery of research and communications skills. Usually, such individuals will contribute to all aspects of the research problems in the laboratory, ranging from choosing new problems worth studying to presenting papers to the thought collective.

The precise way in which the senior investigator of the laboratory interacts with associates varies considerably depending on the particular situation in the laboratory. It is useful to think of a research laboratory as a small business. The senior investigator is the owner/operator who decides whether to keep the business small or to try to expand. The major products of the business are the research contributions. Productivity refers to the number and significance of research papers published by the laboratory. In addition, the laboratory advertises its products through seminars and papers presented at scientific meetings. Finally, the income of the business is derived from research grants and other awards given to the senior investigator; such awards are generally granted on the basis of the thought collective's assessment of the productivity of the investigator.

Each member of the laboratory is an employee of the senior investigator, who provides working space, supply money, and salaries. Even when graduate students or postdoctoral fellows get their salary support from sources other than the senior investigator, they will be dependent on the senior investigator for space and supplies. Because of the limited funds available, the senior investigator must constantly take into account the cost effectiveness of his/her laboratory activities. *Experiments are planned according to whether they are affordable as well as whether they are scientifically worthwhile.* Doing science is not simply a matter of going into a laboratory and carrying out research. In fact, because of the difficulties involved in obtaining funds, some investigators spend as much time writing grant applications as they spend carrying out research.

Since productivity of the laboratory is a critical factor in the ability of the senior investigator to continue to obtain adequate funding, the contributions of technical assistants, graduate students, and postdoctoral fellows are important in terms of the productivity of the laboratory. Nevertheless, the graduate students and postdoctoral fellows in the laboratory have needs other than demonstrating productivity. For instance, it is not enough that graduate students learn the technical skills to do experiments. Just as important is that they learn how to design experiments and to choose new problems. If the senior investigator always tells a student what to do, then the student may be a productive author on many research papers, but without learning the other aspects of doing science. Similarly, when it comes to writing papers, the senior investigator usually can write a paper more quickly than a student or postdoctoral fellow. Letting the student or fellow write the first draft almost always slows down the total time it takes to publish the research. From the point of view of the graduate student or postdoctoral fellow, however, learning how to write papers is an important skill that they must acquire. Such considerations reveal that possible conflicts can occur between the need for laboratory productivity and the needs of students and fellows who are members of the laboratory.

It is instructive to imagine the following situation in a laboratory that is in its fourth year of a five-year grant. During the first three years there has been satisfactory but not outstanding productivity. An advanced graduate student in the laboratory comes up with a novel idea that s/he would like to explore, rather than continuing on the major laboratory project. From the point of view of the student's training, this development should be encouraged. But can the senior investigator afford to let the student move in a new direction? If the student stops working on the major laboratory project, productivity in the fourth year may decline, thereby leading to a loss of or a decrease in funding when the senior investigator goes to renew the grant. As a result, the investigator may discourage the student from moving in new directions in order to maintain at as high a level as possible the productivity of the laboratory.

If the above situation were modified slightly, and the laboratory was in the first year of a five-year grant, the investigator would have much

more flexibility. In this case, s/he might encourage the student to begin
the new project. Not only would this be good for the student's scientific
development, it might also lead eventually to a new direction of research
for the entire laboratory.

EXPANDING THE LABORATORY
THOUGHT STYLE

Despite the effort and commitment required to train graduate students
and postdoctoral fellows, most investigators welcome these scientific
"offspring" to their laboratories. This is the means whereby the continuity of science is established, and there is a certain sense of im-
mortality in passing down one's thought style. In addition, and possibly
of greater importance, the presence of students and fellows expands
the thought style of the laboratory.

The senior investigator's thought style is limited to previous research
training and experience. During an investigator's career it often happens
that s/he must learn new kinds of research methodologies or even
become involved in new research fields in order to continue to be
productive. One must either learn new ways of approaching old problems
or study new problems. Such transitions are very difficult, especially
for older, established scientists. Learning the literature of a new field,
for instance, requires a massive effort. But without such an effort, an
individual's science often becomes "just more of the same." Here is
the situation in which graduate students and postdoctoral fellows often
play a critical role. Through them, the senior investigator can begin
to use new methodological approaches and develop research problems
in new fields of interest.

Graduate students are scientifically adventuresome. This trait man-
ifests itself in several different ways. National scientific meetings are
accompanied by large commercial exhibits. The exhibitors sell the
equipment and reagents necessary for carrying out research, and new
instruments and reagents are introduced at such meetings. It is inevitable
that students will come away from the exhibits with some new technique
that they wish to try themselves. Similarly, when graduate students
attend a seminar at which a novel method has been described, they
often subsequently propose the use of this method in their own research
problem. In both cases, it is the novelty of the method that attracts
the student because s/he is focused to a large extent on learning about
new methodological approaches. The senior investigator may be aware
of these new methods, but ignores them because they do not appear,
at first sight, to be relevant to the research problem under study in
the laboratory. Only after the student tries the method out and produces

new and interesting information does the potential value of the method become obvious to everyone else.

A similar situation can occur regarding the interpretation of experimental findings. The graduate student, whose knowledge of the experimental system is less sophisticated than that of the senior investigator, sometimes suggests explanations that previously were overlooked because they did not fit into the investigator's thought style. It is precisely because the student has a less focused view than that of the senior investigator that s/he can sometimes take a novel approach to interpreting the data.

Finally, graduate students often help develop new methods that the senior investigator wishes to see initiated. Assigning such tasks to students gives them the opportunity to develop research skills in an independent fashion and, at the same time, enhances the laboratory.

Postdoctoral fellows are similar to graduate students in that they, too, bring new ideas and interpretations to the laboratory in addition to their development of new techniques. If, for instance, a senior investigator wishes to do experiments in a new field about which s/he is only marginally familiar, then a postdoctoral fellow whose training is in the new field can be recruited. Such situations often are ideal for both the postdoctoral fellow and the senior investigator. An investigator studying research problems at the cellular level who wants to orient the studies toward some specific disease can recruit a medical resident whose speciality is in the relevant branch of medicine, and who desires to spend some time doing basic research.

In summary, remarkable diversity is possible in individual research laboratories. The senior investigator's thought style will determine the approach to organizing the laboratory; the reasons for which graduate students, postdoctoral fellows, and technical assistants are recruited; and the opportunities they will have. These attitudes will be learned by students in the laboratory as they carry out their thesis research. The degree to which the students acquire the skills necessary to become independent researchers will be determined in part by the independence allowed the students in the laboratory. At one end of the scale, students may be treated as technical assistants throughout their training and as a result become good technicians but lack many other important skills. At the other end, the students may be placed completely on their own from the very beginning and told to develop their own problem and methodological approaches, with the senior investigator present only as a consultant.

If the student succeeds in obtaining the Ph.D. degree and eventually goes on to become an effective independent investigator, the student's advisor will be confirmed in his/her belief that the training method was valid. If the student is not successful, then the advisor may question the training methods used, but more likely will attribute the failure to the student's lack of ability. Similarly, the student may experience the training method as one that worked or decide that success occurred in spite of the kind of training received. Regardless of how a student and advisor evaluate the quality of the training, an investigator's thought style will be subjected to a process of natural selection, since the thought style will be preserved only if the student becomes an independent investigator and passes the thought style on to the next generation of students.

THE PH.D. THESIS

The final stage in a student's training is preparation of the Ph.D. thesis. Usually the student's advisor determines when the time has come for the student to write and present the thesis. If the advisor is oriented toward research productivity, then the student may not be able to finish until sufficient research has been carried out to lead to publication of several papers. Alternatively, the advisor may feel that the research project can be viewed as a model for teaching the student how to do science, and that publication of the results is not as important. The range in opinions regarding this matter is evidenced by the variation in the length of Ph.D. theses—in my experience, from less than 100 to more than 300 pages.

In its classic form, the thesis is divided into sections on history of the problem, methods used, results accomplished, and discussion of the results. Although this format is similar to that of a research paper, the various sections in a thesis usually are written in greater detail than papers and with much more speculation in the discussion section. The argument sometimes presented, therefore, is that the style used to write a traditional thesis is not similar to the style students have to use later on in their careers. An alternative possibility would be for the student to write a brief introduction and attach reprints of published papers to take the place of the traditional thesis.

The latter method would decrease the importance of the graduate program in assessing the student's mastery of the graduate program thought style. Publishing papers requires communicating according to the acceptable thought style of the journal in which the student's research is published, and there are different kinds of journals that have different requirements and expectations regarding what makes a particular manuscript publishable. In addition, the student's work may be more easy or difficult to publish depending upon the advisor's reputation in the field. The work may be quite good even though it has not been published.

Therefore, the traditional thesis presents the graduate program with an opportunity to judge the work on its own merits.

At one time, the thesis defense served the rigorous function of determining whether students received their doctoral degrees. Recently, however, it has become more of a final initiation rite. This rite serves a useful and important purpose in providing confirmation of the student's postdoctoral status. Up to this point, the student will have mastered the thought styles of the graduate program and the individual laboratory in which the thesis research was carried out. Now the student becomes an investigator with a Ph.D. degree, an "official" member of the thought collective of researchers in the field.

Scientific Collectives: Maintaining the Thought Style

5 Chapter 4 introduced the idea of scientific thought collectives and described how the thought style of the collective is transmitted to new members, especially students. The second major activity of the collective is to evaluate the research performed by individual investigators. As already mentioned, an investigator's work will be considered in the context of the prevailing thought style of the collective. Only research that can be assimilated by the thought style will be seen as acceptably scientific. At the same time, however, the thought style is changeable and will accommodate to newly assimilated observations and hypotheses.

The thought collective evaluates its members in a variety of different circumstances. Reflecting on their careers, most researchers recognize that they are judged periodically by their peers. In terms of career advancement, examples of such evaluations include decisions about admitting a prospective student to graduate school, recruiting a post-doctoral fellow as an assistant professor, and promoting an assistant professor to a higher position. Other major decisions concern evaluation of papers submitted for publication and grant proposals submitted for funding. This chapter will begin with a description of academic departments and how they evaluate prospective new faculty members. Then the discussion will focus on research papers and grant applications. Finally, the problem of secrecy and fraud in science will be considered.

EVALUATING PROSPECTIVE FACULTY MEMBERS

Individual investigators in academic institutions usually have a dual involvement in teaching and research. Consequently, depending upon the investigators' skills, the academic departments to which they belong will develop reputations for teaching and/or research excellence. (Similarly, universities are evaluated according to teaching and research criteria.) While the research accomplishments of individual investigators often are known to scientists at other institutions, teaching accomplishments generally are known only locally. Therefore, the national and international reputations of an academic department typically reflect the research performed by investigators working in the department and by students and postdoctoral fellows trained in the department.[1]

Departments seeking to add a young faculty member will look for an individual who fits the teaching and research interests of the department, and who is likely to enhance the departments' reputation. The first step in finding such a person might be to place an advertisement in a journal such as *Science* or *Nature*. An advertisement for an assistant professor probably would read something like the following:

> The Department invites applications for a tenure-track position at the Assistant Professor level. Candidates must hold an M.D. or Ph.D. degree and have at least two years of postdoctoral experience. The successful candidate will be expected to pursue a rigorous research program in modern cell biology and participate in teaching medical and graduate students. Interested individuals should send a copy of their curriculum vitae, statement of future research interests, and names of three references to the Search Committee.

Variations on the above advertisement might contain, on the one hand, a more defined statement of teaching expectations and no request for future research interests; or, on the other hand, no mention of teaching activities but a more explicit statement of research expectations. That is, depending on the institution and the department, the emphasis may be on teaching, research, or a combination of the two. Often a department needs someone qualified to teach a particular course or to carry out research in a particular field.

Since the new faculty member will be expected to interrelate with other members of the department in carrying out research and teaching activities, candidates will be evaluated not only on their specific accomplishments but also on their thought styles. That is, the search committee will look for an individual who already reflects and would fit easily into the prevailing thought style of the department. To illustrate this point, one can imagine the situation of two departments that need someone to help teach cell biology. One department might be located in a small liberal arts college where undergraduate education is its most important activity. Another might be located in a medical school seeking to establish an international reputation for research excellence. In the first case, the department would not want a young investigator committed primarily to research, although the individual's commitment to establishing an international research reputation is precisely what would fit in well with the other department. In some departments, the single most important question asked is, "What is the likelihood that this individual will become a member of the National Academy of Sciences within the next ten years?"

The initial information used by the search committee to evaluate candidates is their curriculum vitae. A candidate's future potential will be projected from his/her past research training and accomplishments. Of particular importance, the curriculum vitae indicates *the candidate's scientific pedigree*—that is, the previous thought collectives in which s/he obtained graduate and postdoctoral training. If the candidate trained in the laboratory of an investigator known to the search committee, then the candidate first will be viewed *according to the advisor's reputation*. It will be presumed that a young investigator trained in a well-respected research laboratory and department will be capable of doing good research, and that the young investigator's thought style will reflect that of his teachers.

The positive effect of lineage on success in science has been well documented.[2] Of the American Noble Laureates studied in 1972 (Table 4.1), about half had teachers who also were Noble Laureates. Baron Rayleigh, who received the Nobel Prize in 1904, had twenty-four scientific children, grandchildren, and great-grandchildren over the next three decades who became Laureates. In the biological sciences there are lineages such as Meyerhoff-Ochoa-Kornberg-Khorana and von Baeyer-Fisher-Warburg-Krebs. The members of the latter group can trace their ancestry back to French chemist Antoine-Laurent Lavoisier. Thus, it is not surprising that the credentials of a candidate will be viewed in the light of the candidate's teachers. It is typically the case that candidates of unknown or questionable pedigrees are subjected to greater scrutiny than young investigators who have more desirable pedigrees. Of course, a good pedigree is no guarantee of success, and there are many young investigators from outstanding laboratories who have not lived up to their advisor's reputations.

In addition to the question of pedigree, the dominant issue that often concerns a search committee is a candidate's publication record. Important factors beyond simply the number of publications are the quality and significance of the work and the contribution of the candidate to that work. To clarify these issues, it is necessary to describe the features and method of evaluation of research papers.

RESEARCH PAPERS

In evaluating research, publications are the singular measure of success. For instance, the number of publications by U.S. scientists and the frequency with which these publications are cited by other scientists are the criteria that the National Science Foundation uses to measure U.S. contributions to world science.[3] In contrast, the number of U.S. patents is used to measure U.S. contributions to world technological advances and innovation. Use of these different measures is a consequence of the difference in motivation underlying basic and applied research. It should be emphasized that this difference is not necessarily one concerning the kind of institution in which the research is being carried out. That is, although academic research typically is carried out in a university setting, it sometimes is performed at research centers that are sponsored by industry. Conversely, applied research usually is performed in industrial research centers but sometimes is carried out in universities.

In applied research, an investigator's accomplishments typically are determined by the products or processes that are developed for commercial use. Novel products or processes usually are patented, and *the motive behind the research activity is to make a profit.* Although papers describing such work often are published, their primary function is to serve as documentation for regulatory agencies (e.g., the Federal Drug Administration or Environmental Protection Agency) responsible for evaluating the safety of the product or process. Such papers do not have as their goal an invitation to others to repeat and expand the work. In fact, critical details may be left out precisely to prevent others from developing similar products or processes.

Basic research, on the other hand (whether it occurs in a university or industrial setting), *is motivated by the goal of expanding the knowledge of the field.* That is, the immediate aim is to know more about the subject under study. The impact of basic research ultimately becomes evident in everyday life by its application and development into products and processes, but this extension of the basic research occurs sometime in the future. The immediate conclusion and only product of every phase of a successful basic research project is publication of the results.[4] This publication is an implicit invitation for others to become acquainted with and use the results. Unpublished results cannot be learned or confirmed by others, and therefore cannot be incorporated into the scientific knowledge possessed by the field. Of course, investigators engaged in academic research also have important personal reasons for publishing papers such as getting a job, getting promoted, building a research empire, or having their names immortalized in the world's literature.[5] The academic saying "publish or perish" is taken for granted by most scientists who realize that their publications will be used as one of the primary indicators of their contributions to a scientific field.

The Structure of Research Papers

A research paper is a formal presentation of new findings in the context of a newly or previously defined problem. The authors have made certain observations that they consider to be of suitable importance to warrant publication, thereby adding their findings to the previous body of scientific knowledge. Moreover, a research paper is an invitation for intersubjective verification by other scientists. The authors assume that other investigators in the field could use the same methods, obtain the same results, and arrive at similar conclusions. Finally, the research paper is a statement of the authors' thought style. The investigators have a certain view of what aspects of a problem need to be studied, what methods of research are appropriate to carry out these studies, and what implication their work has for subsequent research on the problem.

Research papers generally are divided into several sections: Introduction, Methods and Materials, Results, Discussion, and References. Each of these parts fulfills a different function. The Introduction states the current thought style regarding the problem under study. This includes the reasons why the current studies are important for solving the problem, and it connects the work with previous studies of the same problem. In the process, the continuity of scientific knowledge is established.

In the Methods and Materials section, the techniques used in the work are described in sufficient detail that they could be duplicated by others who might wish to confirm the results of the study or carry out related studies. Similarly, the source of materials is specified so that others know how to prepare or where to purchase the supplies necessary for performing the experiments.

The Results section contains a presentation and description of the data collected in the experiments that were performed. The experimental conditions under which the data were obtained are documented in the figure and table legends if this information was not presented in the Methods and Materials. It is of crucial importance to note that only certain types of experiments will be included in the Results section— namely, those that the investigators anticipate will be convincing to others. Many experiments do not work for either technical or theoretical reasons. Moreover, before investigators get an experimental system under control, the results obtained usually are tentative and incomplete. When the system is well understood, however, it will be possible to design experiments that are consistent, complete, and stylistically appropriate for presenting to the thought collective.

In the above context, a distintion should be made between heuristic and demonstrative experiments.[6] Heuristic experiments, which constitute most research activity, are the failures and incomplete experiments through which investigators learn new information. Once the heuristic experiments are completed, it is possible to design demonstrative experiments. These experiments do not extend the investigator's knowledge, but they are necessary for presenting the results to others.

For example, in studies designed to characterize serum factors required for cell growth, an investigator might want to determine a dose-response relationship—that is, an analysis indicating how cell growth (the response) varies with the amount of serum component added (the dose). In such an analysis, the investigator might want to use five to ten different concentrations of serum, but over what range? The ideal answer is over the range from "no response" to "complete response" (i.e., no growth to complete growth). But the investigator does not know initially what this concentration range will be. Therefore, s/he carries out a *preliminary experiment*—that is, a heuristic experiment. After several preliminary experiments, the investigator will have learned precisely the range of serum concentrations over which cell growth responds. This range is the one used in the demonstrative experiment that gets performed for others to see in the published version of the study.

The more time that an investigator spends on demonstrative experiments, the less new information will be learned from heuristic experiments. If, however, the demonstrative experiments are not convincing enough, then the work may not be publishable. Finding an appropriate balance between these two goals is a recurrent difficulty for every investigator.

The results presented in a paper usually are only a small portion of the experiments actually performed. Moreover, the logical explanation for carrying out the experiments described in a paper often is different from what the investigator had in mind at the time the studies were performed. That is, since earlier findings constantly are reinterpreted in the light of later ones, the reasons for having done the older experiments often become irrelevant. The rationale presented in the paper is usually the one currently believed to be true rather than earlier ideas that were discarded later. As a consequence, earlier experiments may be discussed after later ones, and the work will be presented as a self-consistent, logical process quite unlike the historical process that actually occurred. In describing their studies, many investigators take the role of storytellers, and the plot of their stories is a drama of search and discovery.

Finally, the Discussion is an interpretation of the results according to a particular thought style and in the context of the problem put forth in the Introduction. Included in this section is a description of how the results and their interpretation extend the current understanding of the problem. Also included are the future steps necessary for further clarification of the problem, thereby suggesting a future direction for the thought style.

Frequently, the conclusions of a paper are summarized in pictorial models. Such models have both scientific and artistic features. Scientifically, they serve to organize the observations and hypotheses under discussion, and to put them together in a unified manner.[7] If successful, the model not only reinforces the data already collected but also makes obvious what future experiments are needed. Artistically, models range from schematic to realistic depending upon the investigator's aesthetic taste. In this connection, there is the risk that a model will be interpreted too literally, especially if it is being used as a teaching aid for non-scientists.[8] Some investigators avoid this problem by referring to their models as "cartoons."

In summary, the research paper contains two components: the experimental results and the thought style necessary for carrying out the experiments and placing them in context. When reading a paper authored by more than one investigator, the reader cannot be sure who is responsible for the different aspects of the work. It has been suggested that each author of a paper should be responsible for its intellectual content—that is, for both development of the approach and interpretation of the data.[9] This idea, however, neglects the different roles of individuals in a laboratory. Although one might expect a student's advisor to ensure the reliability of the student's observations, it is unlikely that the student would be asked to approve the advisor's thought style. The senior investigator is the dominating influence in the laboratory thought style and also will be the dominating influence regarding the thought style presented in papers published from the laboratory. One way to approach the disparities in seniority between authors is to expect that each author will be prepared to take responsibility to the same degree that s/he is willing to accept credit for the contents of a paper.[10]

Some papers published by a student and advisor specify the student as the first author and the advisor as the second author. In most such cases, it is likely that the student technically carried out the bulk of the research. Beyond that, however, one cannot be sure of the student's contribution. In fact, in the early stages of a student's career, one would expect only a minimal contribution to the thought style behind the work. Nevertheless, not only are students co-authors of the work, but they may also be asked to write the first draft of the paper even though the advisor anticipates its complete revision. This is part of the student's training.

The Evaluation of Research Papers

The first phase in publishing a research paper is concluded once the paper is written. The next step is to select a scientific journal to which the work will be submitted for publication. As will be discussed below, journals vary considerably in prestige and in the thought styles they represent. Consequently, choosing the "right" journal in which to publish a paper is very important since the choice will affect who becomes acquainted with the research and how quickly. Prior to acceptance for publication, research papers usually are reviewed by the journal for scientific style and content. Typically, the journal sends the manuscript to one or several experts in the field for their critical evaluations. These experts, called *referees*, recommend that the paper be accepted, rejected, or revised. Although the referees know the identities of the authors of the work, the identities of the referees are kept secret from the authors. This is one of the few instances in which secrecy in science generally has been accepted.[11]

It has been suggested that manuscripts would be reviewed more fairly if the investigators' names, affiliations, and grant support acknowledgments were removed. In this case, even if referees could make good guesses about the identities of the investigators who performed the studies, they could not be sure. On the other hand, investigators' previous accomplishments are recognized as supporting documentation that add to the credibility of the work presented in a manuscript. In part, the referees have the responsibility to keep the scientific enterprise intellectually honest.[12]

The scientific evaluation is made on the basis of criteria established by the journal. The more prestigious the journal, the more rigorous the evaluation, and the more difficult it is to get a paper published by that journal. While some journals accept most of the papers they receive, others may reject 90% or more. Especially high rejection rates develop as a consequence of a journal's high prestige. That is, the most prestigious journals receive many more papers than they can publish, thus allowing the journal to set even higher standards for acceptance. Moreover, journals are well aware of their status. For instance, through the February 1982 issue, *Cell* described itself as a journal that "publishes research articles and reviews of general interest in biology." Beginning with the March 1982 issue, however, *Cell* described itself as a journal that "publishes reports of *exceptional significance* in any area of biology."

The questions listed below are typical of those asked of referees by journals in the process of reaching a decision regarding a submitted paper.

1. Is the subject matter appropriate to readers of the journal?
2. Are the methods used suitable for the questions under study?
3. Are the data convincing?
4. Do the data warrant the conclusions drawn?
5. Is the work innovative?
6. Does the paper make an important contribution to the field?
7. Will the work be of broad-reaching interest or of interest only to specialists in the field?
8. Is the paper in the top 1%, 5%, 10%, etc., of papers written in this field?

The first question needs little explanation. Different journals cover different subjects, usually listed by the journal in its Instructions to Authors. In this instance, the referee is being asked to confirm that the work described in the paper is consistent with the stated intent of the journal. Questions 2 through 4 concern the scientific quality of the work. Regardless of whether the problem under study is trivial or broad-reaching in its consequences, one can ask whether the experimental design appears appropriate, whether the data are presented adequately, and whether the interpretation of the results is reasonable. In response to these questions, the referee must decide whether the methods and materials are presented clearly enough so that others could repeat the work if they wished.

The referee must also determine if both the results presented and the authors' description of the results are adequate and convincing. Important experimental details may be absent. The authors may say that a certain structure can be seen in a photograph or a certain pattern can be observed in a data graph, but the reasonableness of these claims may not be obvious to the referee. Often referees may wish to know how reproducible the results were—that is, how may times the experiments were performed. And, if a journal has high standards regarding the quality of figures, criticisms may be raised concerning the aesthetics of the figures even if the information presented is not called into question. I have been told by referees, for instance, to produce higher-quality photomicrographs (with more contrast and sharper focus) so that they will *look better* when published in the journal.

Finally, the referees must consider the interpretation of the results. Authors may have overlooked or failed to mention key control experiments in the absence of which the data cannot be adequately interpreted. The referees may also consider the results and their interpretation to be in conflict with other studies already published but not discussed in the article under question. In this case, the authors may be asked to explain the differences, and the paper may be rejected if reasonable explanations cannot be produced. Previously published work has priority. Once the results of a study have been made public, they become part of the potential thinking of the thought collective and can be (and often are) used to evaluate subsequent work in the area.

In all of these considerations, the referees act as representatives of the thought collective. The purpose of their evaluation is to determine whether the work is consistent with the prevailing thought style regarding how experiments should be performed, presented, and interpreted. When they do their jobs properly, the referees can be a significant help to the authors' research. For instance, I have on many occasions modified hypotheses or carried out additional control experiments that were necessary to clarify points noticed by the referees but which I had overlooked.

Questions 5 through 8 go beyond the adequacy of the work and concern its potential impact. More prestigious journals ask these questions because they want to publish research that is most likely to have a significant impact on the field. In this case, the referees act as representatives of the thought style in that they evaluate the importance of the work in the context of the field. On the one hand, the research may be a major contribution to a field itself no longer considered important (though it might have been central ten years ago). Alternatively, the research may be in an important field, but not a major contribution. In either case, the prestigious journal would probably not wish to publish the study. Referees often say that a paper is not novel enough for Journal "A" and suggest that it be sent to journals "B" or "C."

The subjectivity involved in making the above evaluation is considerable because investigators may vary widely in their opinions about what problems need to be solved, even if they agree on the general research techniques that are useful. Competing groups often exist within fields. Each group, itself a thought collective composed of several different laboratories working independently, has its own thought style regarding what constitutes progress in the field. Work from other groups may be viewed as misguided or misleading, and such considerations often have an influence on how referees regard submitted papers,[13] although not always.[14]

Another aspect of a paper's evaluation is the referee's relationship to the authors of the paper. Usually, referees are established investigators who have contributed to the field in the past. Because of a referee's success in making such contributions, s/he is assumed to have helped to define the prevailing thought style. The author of a submitted paper (at least the senior author) may be a newcomer to the field, or an established investigator, or one of the leading authorities in the field. With newcomers, referees sometimes are arbitrary regarding the author's understanding of the thought style; but such is not usually possible with more established investigators. Therefore, political relationships outside of science *per se* can exert an influence. The referee recognizes that s/he may become known to the senior authors and that their positions (author and referee) will undoubtedly be reversed in the future. In addition, a referee who previously accepted the research and views of the authors may be more generous in evaluating their subsequent work. As a consequence, the referee may accept papers from established investigators that would have been rejected had the authors been unknown. Moreover, when a referee does reject an established investigator's paper, the tone of the rejection is likely to be more conciliatory than occurs with unknown investigators.

Listed below are excerpts from reviews of two papers of mine that were rejected by the referees. Review #1 was written in response to a paper submitted soon after I started working in the field of cell adhesion. Review #2 was in response to a paper submitted twelve years later. During the intervening period, my status in the field went from unknown to established investigator, and the tone of the reviews changed accordingly.

A. Referee's comments on Paper #1, which was submitted to a cell biology–oriented journal:

As I see it, under different conditions cells may use different mechanisms for adhering to surfaces. For example, if I were to drop a cell suspension onto a hot stove they [sic] would adhere to it. This information would in itself be of little use. . . . The part of the [manuscript] relating to surface free energy and the like is not acceptable in my opinion.

B. Referee's comments to Paper #2, which was submitted to a biochemistry-oriented journal:

This manuscript extends Dr. Grinnell's pioneering studies on the interactions of cell with substrates. . . . [It] is regrettably not appropriate for publication. . . . It provides some information that will be of interest to the biomaterials field, and possibly to a few workers in the fibronectin field, but it appears to be of little general interest to the . . . readership.

When a paper is rejected, it may be because the work is not worth publishing, because the work is not viewed as novel enough if submitted to a highly prestigious journal, or because the work has been submitted to the wrong thought collective. Regarding my rejected manuscripts, I believed the last to be the case. In confirmation of my belief, both manuscripts subsequently were accepted after evaluation by journals representing other thought collectives. Each journal tends to utilize a different set of referees, who are selected by the editor managing the review of the paper. As a consequence, papers submitted to a second journal are likely to get a different evaluation from that received by the first. The two excerpts below were written by referees who accepted my manuscripts leading to publication of the work. The differences in the responses of the referees to the papers further points up the fact that similar work can be viewed differently on the basis of two different thought styles.

AA. Referee's comments to Paper #1, which was resubmitted to a biochemistry-oriented journal:

I think that this very interesting paper should be published provided that the crucial issue of cell viability can be satisfactorily dealt with.

BB. Referee's comments to Paper #2, which was resubmitted to a cell biology–oriented journal:

This manuscript is carefully documented, and describes a series of physiologically relevant studies in fibronectin. It should be of interest to cell biologists in the areas of cell adhesion and wound healing.

The views of the referees are an important factor, but not the final determinant of a paper's acceptability. The authors have the opportunity to write a response to the referees and can disagree with them. Mediating between the authors and the referee is a managing editor, usually an investigator appointed to the standing editorial board of the journal. Such prestigious appointments, made by the senior editor of the journal or by the editorial committee of a scientific society, usually are reserved for scientists who are leaders in the field. Being appointed to an editorial board is a statement to the investigator that "you are the kind of scientist whose work should be published in this journal."

Given the existence of competing thought styles within a field, journals, too, often compete on the same subject, with different sets of investigators serving on the editorial boards. In fact, journals typically get started because certain investigators (either independent individuals or members of a society) feel that a field or aspect of a field is not suitably covered by existing journals. For instance, when I was a graduate student, we learned that *Biochemistry* was initiated by several senior biochemists who had become dissatisfied with some of the policies of the *Journal of Biological Chemistry*.

The managing editor evaluates submitted papers and selects the individuals who will act as referees. Choosing fair referees is very important—not only because they decide the acceptability of the work, but also because the referees gain an advantage by learning about the research before other investigators in the field are able to do so. Consequently, an investigator may want to prevent certain competitors from evaluating or becoming acquainted with the work and can request that these individuals not be sent the manuscript. After the referees return their reviews of the manuscripts, the managing editor decides whether their comments are reasonable. A preliminary decision regarding suitability for publication is sent to the authors, accompanied by the referees' comments. The authors in turn send their responses to the referees back to the managing editor. Here again, the authors may be either unknown to the managing editor or well-known investigators in the field, and the authors know the identity of the managing editor. The situation can become delicate politically, and the scientific status of the authors inevitably influences the managing editor as s/he considers the referees' comments and authors' responses.

Papers not rejected outright are ultimately often accepted if the authors persevere. A detailed study of the papers submitted and published in the journal *Systematic Zoology* from 1974 through 1979 revealed that, among the submitted manuscripts that never appeared in the journal, only 60% were rejected outright. Forty percent were accepted with revisions but never resubmitted. Almost all of the resubmitted manuscripts were published. Interestingly, those authors who resubmitted and published revised manuscripts were or became recognized authorities in the field, while the others disappeared.[15]

Before moving on to consider the impact of published papers on the thought collective, I believe it is worthwhile to review the problem of a search committee trying to evaluate papers listed on a candidate's curriculum vita. First-author papers will be thought of more favorably than papers on which an individual is listed somewhere in the middle of a string of authors, because it can be assumed that the first author has played a major role in the research—even if s/he is not responsible for the thought style presented in the paper. In addition, papers published in more prestigious journals will be viewed more favorably than papers published in less prestigious ones because the former will be assumed to be of high scientific quality and to have potentially greater impact on the field. The search committee generally ignores the fact that it is easier for a student working with a well-established investigator to get papers published in prestigious journals than it is for an equally competent student working with a relatively unknown investigator. Thus a second bias develops that complements the already-mentioned preference shown to young investigators with favorable pedigrees.

The Impact of Research Papers

Research papers can have different degrees of impact on other investigators. There is an analogy between the spread of scientific ideas and (all negative connotations aside) the spread of disease.[16] In a sense, the observations and hypotheses made by an investigator act like infectious agents that are transmitted to the scientific population by research papers. An investigator whose thought style is susceptible to the idea becomes infected and serves as a carrier who passes the idea on to others. For a paper to have an impact, therefore, three conditions must occur. First, the paper must be read by an investigator. Second, the investigator must consider the research to be meaningful. And, third, the paper must influence the thinking of the investigator—either by confirming what already was suspected or by leading the investigator to think about the problem in a new way.

The first point, that an investigator must read the article, is not a trivial one. In 1982, for instance, the Institute for Scientific Information cataloged 900,000 scientific papers published in 6,600 scientific journals covering all research fields.[17] Within a single field such as cell biology there are large numbers of journals, perhaps 10 to 20 in which 500 to 1,000 papers are published every month.

As no single investigator can read all of the work published in a large field, investigators use highly selective techniques to determine what papers they will read. One way is to have other scientists send copies of their papers. For instance, I occasionally receive requests from investigators asking me to add their names to my "reprint list"—that is, the list of individuals to whom an investigator automatically sends reprints of his/her papers. Of course, many investigators (including myself) do not have such reprint lists. Another method of selecting papers is to subscribe to *Current Contents*, a weekly publication that provides a listing of the tables of contents from many journals published during the previous week. Alternatively, one can use a computer service with access to a publication data base (e.g., *Index Medicus, Science Citation Index, Biological Abstracts*) and select papers according to their subject matter. One of the most common practices is to read through the prestigious journals, on the assumption that the most important work will be published there—a significant reason for which papers published in prestigious journals are more likely to have an impact on the field.

The second point, that an investigator must consider the work to be meaningful, relates to the investigator's view of the thought style contained in the work. If the problem under study is considered uninteresting or the approach to studying the problem is considered to be the "wrong way to go about it," then the study will be discounted. Similarly, one investigator will ignore another's research if the data presented are not convincing, or if the investigator doubts that the data could be obtained by the methods used. In essence, the investigator (for some reason) does not believe that the same results would be obtained if s/he did the experiments.

Finally, the third point is that the work must influence the investigator's thinking (i.e., thought style). It is possible, for instance, that the investigator's response to the paper will be: "I don't know why they did that work; two or three other laboratories already have shown the same thing." Or, "They keep publishing the same thing over and over." Alternatively, the response to the paper might be: "It's good that they confirmed that observation." Or, "I hadn't thought of that before; it's an interesting idea." In the latter instances, the investigator may incorporate the paper into his/her thinking and make use of it in the future. This incorporation will be manifested most obviously when, subsequently, the investigator *refers to the published work.* In the Introduction and Discussion sections of a paper, authors refer to the work of others that they believe is most important in defining the status of the problem under study. In the Methods and Materials section, authors refer to techniques described by others that were chosen to be used for studying the problem. Thus, when a published paper influences an investigator, the consequence will be citation of the paper by the investigator.

In principle, it is possible to determine the impact of a paper by analyzing the frequency with which it is cited. Similarly, it is possible to determine the impact of an investigator on a field by measuring the frequency with which his/her work as a whole is cited. Finally, it is possible to determine the impact of a journal by measuring the frequency with which papers published in the journal are cited. In recent years, the Institute for Scientific Information (ISI) has been analyzing a large number of papers with respect to their citations of other work. In the 900,000 papers recorded in 1982, there were 9.5 million citations. Using the ISI data base, one can determine the frequency with which a particular paper or all of an investigator's work is cited.[18]

Interpreting such citation counts is problematic for several reasons. Comprehensive review articles often are cited in lieu of referencing original research papers, but the reviews may not be written by those who did the original research. Also, without actually reading the papers one may find it difficult to determine whether a paper is important because of its technical content (development of a methodology) or theoretical content (conceptual design and interpretation of results). Finally, although it is feasible to distinguish between citations to first-author papers and citations to all papers of which an investigator is an author, the important feature regarding the role of the investigator in the work is not evident. Despite these limitations, several interesting aspects of citation analysis are worth discussing.

A large proportion (25%) of published papers are never cited at all, not even by the investigators who originally did the work.[19] A possible explanation for this failure to cite the work is that the data or conclusions were quickly realized to be incorrect. Alternatively, there is the possibility that the work is recognized by the authors as being unimportant. Why would investigators publish such work, even if they could convince the referees that it was worth publishing? The answer resides in the fact that the only tangible indication of successfully completed research projects are the papers published describing the results. An investigator who has worked on a problem for a year has nothing to show for that activity unless one or a few publications are produced, even if the work turned out to be uninteresting. Without these publications, the investigator will have difficulty persuading colleagues that s/he is an active researcher; s/he may also have problems convincing granting agencies that the research proposed actually was performed.

A second interesting feature revealed by citation analysis is that some papers have a delayed impact. Citations to most work reach a maximum level two or three years following publication, and then the frequency of citation slowly decreases. Occasionally, however, there are papers that are cited infrequently at first and then more and more so much later when the work is "discovered." When originally published, such studies may not have been consistent with the prevailing thought style either for technical or theoretical reasons. With further progress in the field, however, the thought style changed and eventually other investigators recognized that what originally was believed to be unimportant was in fact a major contribution.[20]

Citation analysis can also be used to study the emergence of thought styles associated with a particular area of research.[21] Such examinations determine the frequency with which previous studies are cited together in subsequent papers.[22] In the initial phases of studies on a problem, several foundational works in the field will often be cited together. As shifts in the thought style change, the original foundational works will be succeeded in importance by other studies. Using such techniques of analysis, a detailed description of the genesis of collagen research has been carried out for the period between 1970 and 1974.[23] In the citations found in work published in 1970, the research emphasis was on the structural aspects of collagens (major connective-tissue proteins), and papers concerning the structure of the collagen alpha chain and its cross-linking were the most highly cited works. The discovery of a precursor form of collagen in 1971 opened up a new area of research that, according to citation analysis, had become dominant by 1973. What occurred during this time was a shift in the thought style. Problems relating to collagen structure, though still important, no longer were seen as *the* central questions to be answered by members of the thought collective engaged in studying collagens.

An analysis of journal citations provides insight into the relationship between journals and fields of study. As already mentioned, different journals represent different thought styles. Investigators in any particular area of research tend to be more familiar with and interested in research published in the journals that cover the thought collectives of which the investigators are members. Therefore, one can analyze the journal origins of citations and determine the journals that represent any particular field.

To illustrate the above point, I selected four journals—two from cell biology and two from immunology. The *Journal of Cell Biology* and the *European Journal of Cell Biology* are the official journals of the American and European cell biology societies, and the *Journal of Immunology* and *Immunology* are the official journals of the American and British immunology societies. The data in Table 5.1 (cell biology journals) and Table 5.2 (immunology journals), taken from *Journal Citation Reports*,[24] show the total number of citations found in the journals in 1982 as well as the 10 journals in which the cited papers most frequently were published. For instance, the papers published in the *Journal of Cell Biology* in 1982 contained citations to 16,899 other papers, and most frequently (2,664 times) these other papers were published in the *Journal of Cell Biology*.

Examination of Table 5.1 reveals a remarkable similarity between the journals most often cited in publications found in the American and European journals that represent "official" cell biology thought

TABLE 5.1

Citations in the *Journal of Cell Biology* and *European Journal of Cell Biology*

Journal of Cell Biology (16,899 citations in 1982)		*European Journal of Cell Biology* (3,260 citations in 1982)	
Journal Cited	Number of Citations	Journal Cited	Number of Citations
J. Cell Biol.	2664	J. Cell Biol.	406
Proc. Nat. Acad. Sci.	1457	Exp. Cell Res.	132
J. Biol. Chem.	1284	Proc. Nat. Acad. Sci.	129
Nature	726	J. Biol. Chem.	119
Cell	668	Biochim. Biophys. Acta	94
Biochim. Biophys. Acta	470	Eur. J. Cell Biol.	94
Exp. Cell Res.	412	Nature	86
Biochemistry (USA)	356	J. Ultrastruc. Res.	65
J. Mol. Biol.	325	J. Cell Sci.	64
Science	286	Cell	57

Source: Journal Citation Reports, 1982, ed. E. Garfield, ISI Press, Philadelphia, Pa.

TABLE 5.2

Citations in the *Journal of Immunology* and *Immunology*

Journal of Immunology (31,002 citations in 1982)		*Immunology* (6,441 citations in 1982)	
Journal Cited	Number of Citations	Journal Cited	Number of Citations
J. Immunol.	6328	J. Immunol.	993
J. Exp. Med.	4245	J. Exp. Med.	708
Nature	1497	Immunol.	443
Proc. Nat. Acad. Sci.	1446	Eur. J. Immunol.	258
Eur. J. Immunol.	1118	Nature	254
Cell. Immunol.	798	Cell. Immunol.	253
J. Clin. Invest.	585	Clinical Exp. Immunol.	158
Immunol. Rev.	581	Proc. Nat. Acad. Sci.	157
J. Biol. Chem.	469	Infect. Immunol.	94
Immunol.	467	J. Immunol. Methods	89

Source: Journal Citation Reports, 1982, ed. E. Garfield, ISI Press, Philadelphia, Pa.

collectives here and abroad (i.e., the cell biology professional societies). Seven of the 10 most-cited journals were found on both lists. The fact that these lists are not normalized to account for the absolute numbers of papers published in each journal may explain why the *European Journal of Cell Biology* failed to appear on the list of journals most often cited in the *Journal of Cell Biology*.

In addition, there was a close correspondence between the journals cited in the American and British journals that represent the "official" immunology thought collective, and seven of the most-cited journals were found on both lists. At the same time, however, Tables 5.1 and 5.2 reveal little overlap. Only *Nature* and *Proceedings of the National Academy of Sciences (USA)*, both multidisciplinary journals, appeared on all four lists of the most-cited journals. These comparisons point up the fact that different thought collectives are represented by different journals that specialize in communicating the research of the thought collectives. Consequently, it is possible to determine the thought collective with which an investigator is allied by noting the journals in which s/he publishes.

Journal Citation Reports can be used for purposes other than defining the journals of a thought collective. It is possible, for instance, to determine the relative impact of different journals within a field by comparing the number of citations received by papers published in the journals. Also, by assessing the average time it takes before papers published in a journal begin to be cited in the literature, one can arrive at a measure of the immediacy of a journal.[25] As would be expected, the most prestigious journals have the highest impact and immediacy.

RESEARCH REPORTS OTHER THAN FORMAL PAPERS

The process of publishing a research paper often takes more than a year between submission of the paper and its appearance in a journal. During this period only the editor and referees assigned to the paper will know of the work under way; but they are not supposed to use the information or to make it known to others. If the investigators wish to make the work known to the collective more rapidly, they must find a means of reporting the work in preliminary form.

Preliminary research reports often are presented at meetings of professional societies. In most cases, investigators request the opportunity to make a presentation at these meetings. By communicating the major conclusions of the studies, they become known to the thought collective, and the investigator (as the first to make these observations) establishes priority. When investigators are invited by the organizers of a professional society meeting to present their current work, this is an indication that their area of research is expected to be of interest to the participants at the meeting. Moreover, the invitation signifies that the investigator is seen as an authority whose recent work probably determines, at least in part, the current thought style of the field.

Investigators also may be invited to give seminars to academic departments (or equivalent groups). Such invitations usually have one of two possible meanings. The first is similar to that described above for invited talks at meetings. That is, the department wishes to know the current state of the field and has invited the investigator as an expert. Alternatively, the invited speaker may be an applicant for a position in the department. Once the search committee screens the initial pool of applicants and selects those candidates that appear to be the most attractive, the next step will be to invite the candidate for an interview that includes presentation of a research seminar. The style and content of the presentation usually will clarify for the search committee and other members of the department the investigator's thought style and particular contributions to the research.

Oral presentations, like research papers, contain primarily demonstrative results. Investigators present those results that they believe are ready for consumption by the thought collective. Unlike research papers, however, oral presentations are reviewed minimally or not at all, and the details of the methods and materials generally are not presented. As a result, other investigators can learn the thought style behind the work, but not the precise way in which the work was accomplished. Moreover, speakers usually can present their research with sincerity and confidence more effectively than writers. Thus, such work may appear very convincing even though documentation of the results would not be sufficient to warrant publication in a journal. When trying to obtain such documentation, one may find that the initial results have been erroneous or misinterpreted; indeed, a significant amount of preliminary research never reaches formal publication. As a consequence, experienced investigators tend to consider the results presented at research meetings and seminars with caution until they are able to examine detailed publications of these results.

Ongoing research studies, too, are discussed at a heuristic level, especially with close colleagues. Groups of investigators from one or several departments may meet regularly for work-in-progress sessions. At such meetings, an investigator discusses the results as they are developing, including experimental failures as well as successes. The group functions as an extension of the investigator's laboratory, helping the investigator carry the work forward, at least intellectually. The members of such groups are "insiders" in relation to the research.

"Outsiders" (such as referees) only get to see and critique the finished demonstration of the results, whereas insiders critique the work as it is in progress.

The continuum that exists between insiders and outsiders takes the form of concentric rings around every research project.[26] The insiders, beginning with the individual laboratory carrying out the studies, are familiar with and can discuss the work in esoteric detail, often using jargon totally unfamiliar to outsiders. The outsiders are less familiar with the details of the work, particular in its heuristic form; rather, they learn only the demonstrated conclusions and implications of the studies. Eventually, the outer circles reach beyond science into everyday life when the potential impact of the work is brought to the attention of the public.

RESEARCH FUNDING

Through its evaluation of research papers, the scientific thought collective determines the acceptability of completed studies. Through its decisions about research funding, the collective determines what studies will be performed in the first place. In 1985, an individual research laboratory consisting of an assistant professor along with a technician, postdoctoral fellow, and graduate student would require at least $100,000 per year in salary support in addition to supplies and equipment. Budgets for individual laboratories well in excess of $150,000 are not uncommon. Even if the institution pays the entire salary of the senior investigator (which is usually not the case), other sources must be found to pay the remainder of the expenses. While some funding is available through private contributions (e.g., endowed chairs) and from industry, most money for biomedical research in the United States is obtained from the federal government through the National Institutes of Health (NIH) and the National Science Foundation (NSF).

An investigator who wishes to apply for research funding from a federal agency begins by preparing a formal proposal. The investigator's institution then submits the proposal to the funding agency—for instance, the NIH. The process used by the NIH to evaluate submitted proposals is called the "peer-review" system, which will be described below. By means of this system, the NIH tries to make fair and efficient evaluations and decides which of the many grant proposals—20,000 in 1984—should be funded.

When a grant proposal arrives at the NIH, it is reviewed by the Division of Research Grants. An administrator reads the abstract and perhaps other parts of the grant and makes two decisions. First, the NIH institute most relevant for the proposal is selected. There are several institutes, each of which is concerned with different aspects of biomedical research. Some are focused (cancer, aging) while others cover a variety of research areas (general medical sciences). Second, the administrator selects the study section that is appropriate for reviewing the application, depending upon the kinds of studies being proposed. Study sections such as "Molecular Cytology," "Cell Biology and Physiology," and "Cell and Molecular Biology" review grant applications in cell biology but specialize in different aspects of the discipline. For instance, a proposal emphasizing the normal aspects of cell structure and function would be directed to a study section different from that receiving applications concerning the structure and function of cells associated with a specific disease state.

The institute and study section assignments are critical. First, institutes vary in terms of the amounts of money they have available for supporting research. Thus, an application of sufficient merit to be funded by one institute might not be fundable if assigned to a different institute. Second, adequate review of the proposal by a study section requires that there be investigators on the study section who are familiar with the field of research and the methods described in the application. It is the study section that "approves" or "disapproves" the grant application, assigns the application a funding priority score, and determines whether the proposed time and budget for the research are consistent with the studies described in the application.

Finally, reviewed proposals are sent to the institute's advisory council. By and large, this council ratifies the work of the study sections and awards the available funds to the approved grants in order of the priorities recommended. Sometimes the advisory council may decide to fund a proposal even when the priority score assigned by the study section would not otherwise be considered good enough. The latter might occur, for instance, if the goals of the grant proposal were considered by the advisory council to be of especially high priority.

Most NIH study sections meet three times a year. Between 12 and 18 scientists constitute the membership in each. These individuals are well-established investigators—respected in their fields and of the same stature as those investigators generally asked to serve on journal editorial boards. An investigator whose grant applications have been favorably regarded by a study section may subsequently be asked to join the study section, as such an individual has demonstrated a thought style that is consonant with that of the study section. In short, the study section tends to conserve its thought style.

During the study section meeting a large number of grants (often 80 or more) are reviewed. Prior to the meeting, grants assigned to the study section are carefully evaluated by the members, at least two of whom are assigned to be the principle reviewers. They write critiques of the proposed research, investigator, and budget, and generate a recommendation regarding funding priority. These critiques are read to the entire study section at the meeting. Other members of the study section are invited to comment if they wish. If there is some disagreement, an attempt is made to clarify the diverse opinions and reach a consensus. Finally, the entire study section votes a priority score. Then the budget is discussed and adjusted if necessary.

The discussion of each grant takes only about 15 minutes, but it is usually possible to reach a consensus in that time. The extent of agreement between the individual evaluations of a grant by the two primary reviewers is generally remarkably high. One aspect of the rating system, however, is arbitrary. *Priority scores of approved grants are assigned on a continuum between 1.0 (best) and 5.0 (worst), but funding of grants is either complete or not at all* depending upon where the funding cut-off is set. For example, in an NIH institute with a 1.70 funding cut-off score, a priority score of 1.68 will be funded whereas a grant with a priority score of 1.72 will not be. Study section members will readily admit, however, that it is almost impossible to distinguish between a grant worth a score of 1.65 and another worth 1.75.[27]

Serious consequences result when a grant is not funded. In the first place, the unfunded investigator may have to stop or curtail research activities until another grant aplication or a revised version of the original application is funded. As a result, the research staff in the laboratory may lose their jobs. In addition, since most investigators receive some or even most of their salaries from grants, the position of the investigator may be in jeopardy. Also, the research proposed by the investigator will not be carried out, even though it was viewed favorably by the study section. Of particular importance, the investigator may become discouraged by the system, and choose not to continue doing research. Thus, membership in the scientific thought collective will be diminished.[28]

The Structure of Grant Proposals

Research grant applications contain three major sections. In one section is a statement of how much the proposed research will cost; in another, a description of the investigator's background and previous research activities; and in the last, a description of the proposed research to be performed. The description of proposed research begins with the general problem to be studied (i.e., the long-term goal). Investigators present their views on what hypotheses have been well established in the field and what lines of research should be followed next. A rationale linking the past and future should establish a conceptual framework for the research. In addition, it is important that the studies be placed in a broader biomedical context. That is, the relevance of the research to health-related problems needs to be addressed. Will the research, for instance, help understand or cure some disease of clinical relevance? After the investigator has described the general problem to be studied, the specific aims of the proposal are explicitly stated. These consist of the important questions that the investigator intends to answer during the period of the grant. Usually, the research is projected over a period of two to five years.

The proposal goes on to present an experimental plan describing how the specific aims will be accomplished. This plan usually is the most detailed and lengthy part of the proposal. The rationale for each experiment to be carried out is described, and the technical details of how the experiments will be carried out are specified. In addition, the investigator indicates how s/he anticipates the experiments will turn out, and how the anticipated results will be interpreted. Finally, the investigator states what difficulties might be encountered, either technically or in interpretation of the anticipated results, and indicates the appropriate control experiments and alternative approaches that will be used to help deal with these potential problems.

Often included in the experimental plan is a section on preliminary results. It is in the investigator's interest to document that the proposed experiments are feasible, and that s/he has the facilities and experience to carry them out. Alternatively, an application for the renewal of a previously funded grant will contain a section on previous progress. Future experiments proposed may be seen by the study section as unlikely to succeed if the investigator has not demonstrated substantial progress in the past.

A research grant proposal thus details all aspects of an investigator's thought style, including beliefs regarding the state of the field, future directions, and how future studies should be carried out. Many of an investigator's hidden assumptions are revealed, particularly when the investigator describes his/her expectations regarding the outcome of the experiments and the necessary controls.

The Evaluation of Grant Proposals

Study section members are given certain instructions for evaluating grant applications. They are told, for instance, that budgetary considerations should not affect the priority score, which instead should be determined solely in terms of the scientific content of the application and the investigator's qualifications to carry out the work. In other words, the priority score is voted by the study section before the budget is discussed. In addition, the study section members are directed to assign priority scores without trying to "second-guess" the level at which the institutes will be funding. That is, the full range of scores should be used, and grants should not be given better priority scores because the funding level cut-off has been lowered.

While most reviewers are able to separate the quality of research from its projected cost, assigning priority scores independent of anticipated funding levels turns out to be very difficult. In the jargon of study sections, reviewers are asked to indicate their "enthusiasm" for a grant application. A reviewer who has very high enthusiasm for a grant application *wants the grant to be funded* and will therefore generally recommend that the study section assign the grant a priority score in the fundable range.

The fundable range has been shrinking, however. In 1978, one of my grants was funded with a 2.32 priority score. In 1981, a requested renewal of the grant was approved but not funded with a 1.82 priority score. A revised version of the application was resubmitted and funded after receiving a priority score of 1.62. Then, in 1986, I had another grant approved with a priority score of 1.61, but funding was delayed. Study sections that in 1978 could comfortably assign a grant a priority score of 2.0 or better and assume that the grant would be funded,[29] by 1986 were assigning a grant score of 2.0 if they wanted to make sure that it did not get funded.

A number of different factors influence the referees as they evaluate a grant application. In a 1981 survey carried out by NIH, 44 different types of defects were noted in applications. Some of these defects are listed in Table 5.3.[30] The first class concerns the general nature of the proposal. Sometimes the proposal was poorly written and the referees did not understand what the investigator had in mind. In other instances, the referees disagreed with the importance of the work or with the relevance of the questions proposed for study. Work not seen as being important and meaningful, even if successful, will be anticipated to have negligible impact. In still other instances, although the field may have been an important one, the referees felt that nothing new was suggested in the proposal. The questions raised already had been answered. In short, the referees' and investigator's thought styles were in opposition. The two groups disagreed on both the status of the field and the direction in which it should be moving.

Because conflicting thought styles lead to decreased enthusiasm for a grant proposal, it is essential that the reviewers of grants be familiar with the field in which the proposal is written. Otherwise, the reviewers may make substantial errors, and the proposal will not be appreciated

TABLE 5.3
Common Defects in Grant Proposals

Defect	Frequency (%)
I. Defects in the general nature of the research problem selected:	14.2
A. Experimental purpose or hypothesis is vague.	
B. The problem is of insufficient importance or biologically irrelevant.	
C. Proposal is repetitive of previous work.	
II. Defects in experimental approach:	59.1
A. The overall design is unsound, or some techniques are unrealistic.	
B. Proposal is not explicit enough, lacks detail, or is too vague or general.	
C. Some problems are not realized or dealt with adequately.	
D. Methods or scientific procedures are unsuited to stated objectives.	
E. Application is poorly prepared or poorly formulated.	
F. Assumptions are questionable; evidence for procedures is questionable.	
G. The results will be confusing, difficult to interpret, or meaningless.	
H. The design is too ambitious or otherwise inappropriate.	
I. The approach lacks scientific imagination.	
J. The approach is not rigorous enough, or it is too naive or too uncritical.	
III. Defects in the investigator's scientific credentials:	20.3
A. The investigator does not have adequate experience for this research.	
B. The investigator's knowledge or judgment regarding the literature is poor.	
C. The results from the previous year's support are inadequate.	
IV. Other	5.4

Adapted from: Why grant proposals are unsuccessful. N.I.H. Health Grants and Contracts Weekly, March 31, 1981, Bethesda, Md.

even if it is sound. Moreover, in the research grant application process, unlike the reviewing process for scientific papers, the investigator has only a limited chance to present counterarguments to the reviewers. Therefore, an experienced investigator looks up the members of the study section to which the grant is assigned by the Division of Research Grants and can request a change if s/he feels that the field of interest is not adequately represented.

The second class of defects in grant proposals, and the area in which most "errors" are made, concerns the experimental procedures. The referees may have believed that the experiments would not work as described and should have been formulated in a different manner. Or they may have believed that the experiments were do-able but that the results would not be interpretable. In this case, the reviewers disagreed with the set of assumptions made by the investigator. Thus, even if the reviewers agree with the investigator on what research should be done, they may have markedly different thought styles regarding how the research should be done.

Taken together, the first and second classes of defects reveal the possibility that reviewers will consider unimaginative grant applications trivial and highly innovative applications fanciful. Compounding this situation is the requirement that the investigator project the anticipated work over a period of three to five years. The art of grantsmanship is to create a proposal that appears to be do-able and worth doing but not already done. A commonplace tactic used as a part of grantsmanship is to propose studies already accomplished in the laboratory but not yet published.

In the final analysis, the reviewers are forced to deal with the grant proposal as a model of the investigator's thought style. The studies planned indicate what kinds of questions the investigator would like to ask and how they will be asked, but the research may occur in a way different from that proposed. Generally, it is recognized that the investigator will probably not carry out the research precisely as it is described in the grant application. One cannot plan three to five years (or even three to five months) of research in advance when one is investigating the unknown.

The above features raise the concern that the structure of grant proposals and their evaluation are contrary to the goal of innovative discovery.[31] Proposing novel studies on poorly understood phenomena probably is not the best way to get funded. Since one cannot predict the future, research support for established investigators might be based on philosophy and previous accomplishments rather than on specific aims and technical details.

The last major class of defects in a grant proposal concerns the investigator. In the section on the investigator's background, the reviewers learn of the investigator's current and previous positions, education, and papers published. That is, the reviewers know the investigator's pedigree and previous productivity. If they feel that the investigator's background is unsuitable for carrying out the research, or that the investigator's productivity has been inadequate, then the reviewers may anticipate that the investigator will be unlikely to succeed even if the proposal otherwise appears to be a good one. Conversely, if the reviewers are familiar with the work of an investigator and recognize the individual as an important contributor to the thought style of the field, then certain flaws in the experimental procedures might be viewed with less concern.

Because an investigator's pedigree and productivity are important features in a reviewer's evaluation of a grant, one might anticipate that young, inexperienced investigators would be at a disadvantage compared to others who are well established. Such, however, is not the case. In recognition of the need to get young scientists started as independent researchers, there are special grants for new investigators in which the study section is concerned with the individual's promise for making future contributions. Often, it is easier to get a first grant than it is to get a first renewal.

SETTING THE AGENDA
FOR FUTURE RESEARCH

It should be evident from the preceding discussion that an investigator's thought style is judged by representatives of scientific thought collectives every time a grant application is submitted for funding or a research paper is submitted for publication. The representatives of thought collectives who make these decisions set the course for future research. Each grant proposal that is funded will potentially lead to novel experiments and new information. Each paper that is accepted for publication contains findings that may advance the field and suggest new questions that need to be resolved. In either case, the experiments proposed or accomplished will apear as a natural extension of what is already known (the prevailing thought style). Otherwise, it would not be possible for the thought style to assimilate the new information and ideas and accommodate to them accordingly.

For most investigators, the pressure applied by the thought collective to comply with the prevailing thought style is irresistible. While individuals occasionally turn up who are able to continue to do science outside of existing thought styles, their work is often without impact or the impact is delayed. They will have difficulties finding research funding and publishing their results. The natural tendency among investigators, then, is to find a niche, a thought collective in which they fit. As already indicated, papers that are unacceptable to one journal may be suitable for another. Similarly, grant applications that one study section finds uninteresting and commonplace may appear exciting and innovative to another.

The agenda for future science is also set by the activities of formal scientific thought collectives (i.e., the professional societies). For instance, research in cell biology was encouraged by the American Society of Biological Chemists when they informed their membership on November 26, 1986, that the definition of a biochemist had been changed from "anyone devoted to the investigation of the molecular basis of life processes" to "anyone devoted to the investigation of the molecular *and cellular* basis of life processes" (emphasis in the original).[32]

The meetings of professional societies exert a major influence on future research through the topics selected for invited lectures and symposia. Individual investigators attending the meetings have an opportunity to learn about these subjects and become interested in them. The subjects selected typically concern very active areas of research in which there is a lot of current interest. Conversely, topics not selected often are of less interest at the moment; either little new information on the subject has been forthcoming in the recent past, or not enough is yet known about the topic to warrant a presentation. Those subjects that are viewed as important attract investigators because more funds are usually available to perform these studies, and the results are publishable more readily than new information regarding a topic not of general interest. The presence of more investigators working within a particular area of research leads to faster development in the area, while progress will be slower if there are fewer investigators working in the field.

THE SCIENTIFIC ESTABLISHMENT

Members of editorial boards and study sections officially represent the thought styles of journals and granting agencies. Officers of scientific societies are official representatives of the society's thought style. The individual investigators who receive recognition for their scientific contributions and become representatives of their thought collectives can be considered part of an informal scientific establishment (the power structure of the thought collective). At the top of the power structure are the scientific elite—that is, the members of the National Academy of Sciences (e.g., in the United States), the Noble Laureates, and other major prize winners. The work of these investigators typifies what the collective views as the model of scientific research.

Investigators become part of the scientific establishment largely through their individual accomplishments, and their individual impact on a given scientific thought style can be verified through citation analysis.[33] Nevertheless, there are many investigators whose work, according to citation analysis, has made a major impact on the thought styles of their fields but who have not become members of the establishment. One has the impression that it is easier to become part of the establishment if an investigator has the right scientific pedigree—in particular, if as a graduate student or postdoctoral fellow s/he worked for investigators who were part of the establishment. One explanation for this situation is that the best possible training is available in the laboratories of such investigators. Another possibility is that an investigator will find it easier to become part of the scientific establishment if that individual is known personally by others who already are part of the establishment. In recognition of this situation, investigators acknowledge that they sometimes attend meetings precisely in order to get to know and become known by others in their field.

A second factor influencing which investigators become part of the scientific establishment concerns the politics of universities and academic departments. The national and international reputations of these collectives—reputations based primarily on excellence in research—depend on the scientific status of the faculty members in question. In order to acquire and maintain high status, departments attempt to recruit faculty members who are already part of, or will become part of, the scientific establishment. Given the opportunity, investigators who are part of the establishment are likely to promote the interests and status of young investigators who are in the same department or institution. Thus, the scientific establishment has a tendency to become ingrown (indeed, it is often described as an "old boys' network"). This conservative aspect of science is a mechanism by which the prevailing scientific thought style is preserved (some would say retarded). Recruitment of younger, less well-established investigators to participate in the publication and grant review processes can help break up the old boys' network. One consequence of such recruitment, however, may be a decreased appreciation of the historical development of research fields and an increased focus on research problems and methods that recently have become popular.

SECRECY IN SCIENCE

The mechanism by which thought collectives review submitted papers and grant applications has the inevitable consequence that an investigator's findings and ideas become known to the reviewers well before they become part of the general knowledge of the thought collective. While there is a moral imperative against the use of this privileged information by the reviewers, investigators have little control over the situation. Similarly, if an investigator reveals his/her plans or findings to colleagues prior to their formal presentation, then that investigator has little recourse if those colleagues choose to use the information in their own work.

The critical issue is one of priority. The investigator who is the first to bring new and important findings to the thought collective usually gets recognition for the discovery. In the case of discoveries of major importance, the investigator may win a prize for making an outstanding scientific contribution. More typically, the work is simply cited by others who come to associate the new developments with the investigator. If the study section reviewing the investigator's grant application is aware of an investigator's contributions, the chances that the research will be funded are greater. Moreover, if the investigator is viewed by the thought collective as an authority in the field, then s/he may be asked to represent the thought style in reviewing papers and grant applications submitted by others.

An investigator in academic research has two possible responses to the above situation. One is an attitude of openness, a belief that science progresses most rapidly under conditions of free communication among investigators. Alternatively, some investigators may adopt an attitude of secrecy about their research. They give away no information to colleagues prior to formal publication of their work[34] and, in fact, may even keep some important details from the referees of their papers.

I heard a story about a paper concerning the molecular aspects of a particular gene that, when submitted for publication, purposefully contained misinformation in the Results section. The paper was completely plausible, and unless the referees had repeated the study exactly, they would have had no way of finding out that something was wrong. After the paper was accepted for publication, the journal sent the page proofs to the authors to be checked for typographical errors made during typesetting, at which time the misinformation was corrected. As a consequence, the reviewers of the paper did not learn the actual findings of the work until they appeared in the published version. If the reviewers had attempted to use the information in the version of the manuscript that they had evaluated, they would have been misled. The above trick put the reviewers and the rest of the thought collective on the same basis with respect to knowledge of the work. If, however, the page proofs had not been corrected for some reason, then the misinformation would have become part of the scientific literature.

An investigator can be secretive also by omitting some critical detail from the experimental procedures. In doing so, the investigator may be able to maintain exclusivity and produce new and interesting information without competition from others who wish to use the same system. Often, the omission will not be obvious to reviewers, and the problem will not manifest itself until others try to reproduce the work. Even then, there is no way to know whether the necessary information was omitted inadvertently or deliberately.

The presence of secrecy in science is commonplace in industrial settings, where the accepted motive is profit rather than the advancement of science. When secrecy becomes part of academic research, it is because the investigator comes to view the work as "mine" (i.e., belonging to the individual) rather than as "ours" (i.e., belonging to the collective).[35] After all, my research is what makes it possible for me to keep *my job* and support *my family*. Nevertheless, while the individual's situation may be enhanced by secrecy, the consequence for science as a whole is detrimental. Since the collective lacks free access to the new information, it takes a longer period for the information to be verified and extended by others. When everyone starts keeping secrets, science stops.

FRAUD IN SCIENCE

Even more detrimental to science than secrecy are the consequences of fraud. When fraud is detected, everyone associated with the situation is discredited, and the resulting confusion in the literature and wasted time and effort undermine the normal functions of the scientific collective. Nevertheless, fraud in science is not a new problem. A recent historical account of this subject raises questions about the research of many illustrious investigators, including Galileo and Newton.[36]

Cambridge mathematician Charles Babbage discussed dishonesty in his book entitled "Reflections on the Decline of Science in England," originally published in 1830.[37] He described three types of fraud: trimming (dropping high and low data points that increase the apparent error of measurement); cooking (selecting only the data that fit the hypothesis); and forging (inventing data). In each case, the investigator "improves" the results in order to convince the scientific collective of the validity of the observations and hypotheses. One hundred and fifty years later, many believe that the problem of fraud has become much more serious because science has become "too competitive, too big, too entrepreneurial, and too much bent on winning."[38] An understanding of science as simply another activity in which people are engaged, however, helps explain why fraud occurs. As has been pointed out, "the step from greed to fraud is as small in science as in other walks of life."[39]

Recent cases of fraud that were discovered and well publicized indicate many different presumptive motivations.[40] For instance, Elias Alsabti wanted to improve his vita quickly, so he allegedly took papers published in the scientific literature, retyped them indicating himself as senior author along with fictitious co-authors, and submitted the "revised" versions to obscure journals. William Summerlin supposedly made important observations in his laboratory in Minnesota that he could not repeat after moving to New York, so, under pressure, he falsified subsequent data. When Vijay Soman learned inadvertently that he was "scooped" by another investigator, he allegedly tried to delay publication of her work while publishing his own made-up data, which reached the same conclusions as those of his competitor. Finally, Sir Cyril Burt held such strong beliefs that he apparently made up and published data to convince others that his ideas were correct.

Some of these recent cases of fraud involved young scientists or students who were working in laboratories run by well-known and respected scientists. The question has been raised whether the incidents were investigated promptly and thoroughly at the universities in which they occurred.[41] Although it is essential that such incidents be fully explored, caution is required because of the consequences of improper accusations. For instance, a junior faculty member who was not promoted moved to a new position and then reported to the FBI that his previous department chairman had misused federal funds. The consequence was a full-scale NIH investigation that wasted six weeks of everyone's time and effort before the chairman was exonerated completely.

One difficulty in detecting fraud proceeds from the ease with which an investigator can make up or modify many types of data. Moreover, many scientists have difficulty in believing that fraud has occurred because it seems so purposeless. If the observations are unimportant, why would anyone bother to make them up? Investigators do not become part of the scientific elite by publishing obscure work in obscure journals. If the results are important, then others will try to use the observations. In most cases, the fraudulent scientist will either be found out or end up being viewed as someone whose results cannot be trusted.

One final consequence of fraud is that it undermines the credibility of scientists and the scientific enterprise viewed by those outside of science. This is a matter of major importance because the normal functioning of the scientific collective depends in part on its interactions with collectives outside of science. These interactions are described in the next chapter.

Science
and the World

6 People who are scientists do not spend all of their time doing science. Moreover, when they are doing science, investigators cannot simply forget their previous experiences and interests outside science. On the other hand, nonscientists hear about and read about what scientists are doing, and they experience the impact of science in everyday life. This chapter is about the interactions between science and the world outside science. First, the relationship between scientific observations and routine experience is described. Then the discussion focuses on the effects of science—the effects of its products and thought styles—on everyday life. Also considered are the effects of everyday life on doing science, with emphasis placed on the impact of politics and religion. Finally, there is a brief consideration of the ways in which nonscientists learn about science and the activities of scientists.

THE ORIGINS OF SCIENCE
IN ROUTINE EXPERIENCE

In trying to understand the interaction of science with the world, one must remember that science begins with daily life. As emphasized in Chapter 2, observer and observed form an integrated unit. Through application of the scientific attitude (the thought styles of scientific collectives), the scientific domain of the world is constructed. The scientific domain changes as the scientific attitude undergoes alterations, and the reality described by science is neither complete nor absolute. Yet despite these limitations, science has led to modern technology, which in turn has been influential in changing the world during the past 400 years. Herein resides the "practical truth" of science.

The fact that the scientific domain is derived from daily life appears paradoxical given the apparent contradictions between reality as viewed by the scientific attitude and reality as experienced according to everyday common sense. We have learned from science that the universe—all that exists—is expanding. Based on our common experience, however, we have come to believe that for something to expand, there must be a potential space into which it can expand. Also, we have learned from science that objects such as tables are composed of small particles and contain mostly empty space. But common sense indicates that tables are solid. Perhaps it is not surprising that there still exists an organization called the International Flat Earth Research Society, dedicated to the common-sense belief that the earth is flat and stationary. Research supported by this society is published in the *Flat Earth News*.[1]

Sometimes it seems as if there is more than one reality, one described by the scientific attitude and another described by common sense. One possibility is that the nonobservable objects of science (e.g., atomic and subatomic particles) are somehow different from observable objects (e.g., tables and airplanes). It has been pointed out, however, that objects we routinely call "observable" also are seen indirectly—that is, through air, windows, eye-glasses, and so forth. Consequently, it has been suggested that the distinction between observable and nonobservable objects is accidental, a function of an observer's physiological makeup and the instrumentation s/he has available.[2] Scientists use their instruments to "read" the "text" of nature, and there is a reciprocal relationship between the "nonobservable objects" seen and the instruments used to see them.[3] As long as the existence of such objects makes sense in the context of an investigator's expectations, the reliability of the instrumentation as a means of seeing the objects is taken for granted.

Apparent differences between common-sense experiences and scientific descriptions of these experiences may occur because of the limitations of descriptive language. Science, like other activities of daily life, depends upon human language as derived from typical human experience. As already pointed out, practical concepts necessary for observing and thinking about things are derived from perceptual experiences. The paradox of modern science is that many concepts that need to be discussed refer to aspects of reality beyond the possibility of everyday experience. This situation presents a severe difficulty for those doing science because they are confronted with the problem of trying to imagine what ordinarily is unimaginable.[4]

Trying to talk about some scientific ideas is like trying to explain to a color-blind person what it is like to see red color. The best that can be done is to say that the effect of red color usually is experienced when light of a certain wavelength interacts with an observer's eye. That is not the same thing as seeing something red. Similarly, scientific objects that go beyond common human experience can be accurately described by concepts using mathematical models, but such descriptions cannot be articulated in everyday language (except insofar as one can describe the experimental process by which the objects were discovered).

The most obvious problems relating to the tension between scientific and common-sense views of reality tend to occur in physics. Questions asked by biologists generally do not concern objects that are as large (stars), as small (electrons), or as fast (light) as the objects of interest in physics. The language of biology does present a problem, however, inasmuch as one of the features that makes organisms interesting is their history. For instance, in order to describe the evolving characteristics of a group of organisms that constitutes a unique species, or the developmental relationship between the cellular genotype (content of nucleic acid) and phenotype (expression of particular morphological and biochemical characteristics), one must first understand the present status of the system *and* the historical path by which the system reached this status.[5] One can think of species, for instance, as being characterized by a dimension of time as well as by the three dimensions of space—a difficult concept for many people. Try looking at another person as a historical sequence integrated across time rather than as the individual who confronts one here and now.

EXPERIENCE BEYOND SCIENCE

There is a story about a night watchman who found a man searching under a street lamp for his lost keys and offered to help. When their combined efforts were unsuccessful, the watchman asked the man whether he was sure that he had lost his keys just under the light, to which the man replied, "No, but I can see better here."[6] The looker could not search where he could not see. In a related example, Amtrak trains running between Boston and Washington pass through a number of cities but stop at only a few stations.[7] The traveler cannot visit where s/he cannot disembark. Similarly, the assumptions of science act as constraints that determine what aspects of reality can be investigated scientifically. *Since intersubjective verification is the criterion assumed by scientists to establish truth in science, whatever cannot be studied or verified intersubjectively cannot be incorporated into the scientific domain.*

To emphasize this point, it is worth mentioning some examples of life experiences that cannot be assimilated by the scientific attitude because they cannot be intersubjectively verified by scientists. The following short story is an example of a Zen *koan*. This one has been used by students of Zen Buddhism since the sixth century A.D. as a way to increase their experiential awareness:

> The wind was flapping a temple flag. Two monks were arguing about it. One said the flag was moving; the other said the wind was moving. Arguing back and forth they could come to no agreement. The Sixth Patriarch said, "It is neither the wind nor the flag that is moving. It is your mind that is moving.[8]

A commentary on this *koan* was added in the thirteenth century:

> The wind moves, the flag moves, the mind moves:
> All of them missed it.
> Though he knows how to open his mouth,
> He does not see he was caught by words.[9]

According to this commentary, everything moves, and nothing moves, and the truth of Zen is lost once it has been articulated—that is, "caught by words." The goal in Zen is to experience everyday life *directly*. Unlike the scientific attitude, which depends on observations made in the context of intersubjectively shared, practical concepts (discussed in Chapter 2), the attitude of Zen aims toward experience prior to practical concepts. Since this reality cannot be articulated, it cannot be verified intersubjectively and remains inaccessible to the scientific attitude.

Like the enlightened experiences of Zen, mystical encounters with "God" also are inaccessible to the scientific attitude. The content and meaning of such experiences has been described in written records such as the Bible and similar sources. Most individuals who claim to have had mystical experiences say that the content of the experience is ineffable; that is, the experience itself cannot be described. Nevertheless, the individual believes that s/he has learned about the "true" nature of reality through the experience.[10] Science is unable to study such experiences because of their exclusivity. They are "mine," not "ours." One can believe that they are true, but they cannot be verified intersubjectively. Indeed, the "truths" reported by different mystics based on their experiences have often been contradictory.

Usually, two features of experiences—how typical they are and how clearly they can be described—determine a continuum of what can and cannot be verified intersubjectively. Experiences that typically can be had by anyone and that can be described readily are the ones most easily incorporated into the scientific domain through the scientific attitude. But if Zen masters and religious mystics were able to make practical predictions about everyday life that appeared to be as accurate as those made by scientists, one suspects that the "methods" of Zen and religious mysticism would become part of the scientific thought style before too long.

THE EXPERIENCE OF HEALTH

An aspect of everyday experience of considerable interest to biomedical science is a person's health. Yet health involves individual feelings that often cannot be described clearly. A useful distinction can be made between an individual's condition as determined by analytical measurements (pulse, blood pressure, temperature, etc.) and as determined by a person's feelings.[11] One may be unwell based on blood pressure measurements but feel healthy. Conversely, one may be healthy according to analytical measurements but feel unwell.

From the point of view of the scientific attitude, unambiguous diagnosis of particular disease states depends upon the clear articulation of symptoms by the patient. The statement "I don't feel well, but I can't tell exactly why," often the starting point in patient-physician interactions, cannot be verified intersubjectively. The physician can run a large series of analytical tests hoping to find some aspect of the patient's physiology that is out of the normal range and might account for the unwell feeling. Alternatively, s/he may tell the patient to "take two aspirins and go to bed" in the hopes that either the patient's feeling will go away or definable symptoms will emerge. In either case, the patient may resent the inadequacy of modern, scientific medicine without realizing that the scientific attitude applied to the practice of medicine is in principle incapable of dealing with a patient's feelings if the source of the feelings cannot be articulated.

In general, scientific statements about individual people are statistical; for example, smokers are much more likely to develop lung cancer than nonsmokers, or 40-year-old women have a higher incidence of giving birth to babies with Down's syndrome than 30-year-old women. Before the lung cancer develops or the pregnancy occurs, however, it is not possible to predict with any certainty whether a particular person will be one of those affected. While the scientific practice of medicine has made great advances in diagnosing and treating a large number of diseases whose symptoms can be clearly defined, implicit in this diagnosis is *the view of patients as typical examples of organisms in physiological disequilibrium*. Lost from this view are the individual and unique feelings of patients. Outside of science, the practice of medicine based on faith healing deals with features such as a person's individuality and feelings, but it ignores the intersubjectively verifiable aspects of the patient's physiology.

PRACTICAL EFFECTS OF SCIENCE

The practical effects of science on the world are derived from applied research. In basic research the work is carried out without specific applications in mind, whereas applied research attempts to use the observations and hypotheses of basic research to meet the specific needs felt in the world. Papers published as a result of basic research have no immediate practical effect on daily life. It is only when basic research is developed into products and processes affecting life activities that the practical impact of that research can be felt outside of science.

The magnitude of the effect of basic research often is not recognized unless the connections between the research and its application are direct. For instance, the discovery of a virus associated with a particular disease (basic research) makes possible the development of a vaccine to treat or prevent the disease (applied research). Just as important as the health benefits are the uses of basic research findings in applications originally unanticipated. For instance, the NIH has selected and described ten examples of basic research studies that were used in unanticipated applications.[12] In 1980, these applications were estimated to have contributed $37 billion to the U.S. gross national product—a quantity greater than the total amount of money that was allocated to the NIH by the federal government between 1937 and 1980, and ten times greater than the 1980 NIH budget. For example, enzymologists have been interested in understanding the mechanisms of enzyme action. To carry out such studies, methods were developed for purification of diverse enzymes. Subsequently, the purified enzymes were used extensively by the food and clothing industries in a variety of commercial applications. Individuals outside of science who oppose basic research when they do not understand its immediate relevance fail to recognize science as a body of interrelated knowledge available to everyone, the potential applications of which cannot be predicted in advance.

SCIENCE AND RELIGION

The highly visible effects of science in everyday life have resulted in general acceptance of the scientific attitude as a way of understanding natural events. This acceptance of the scientific attitude has had a practical impact inasmuch as the scientific attitude proposes a model of reality different from other attitudes toward experience, especially those of some religious groups. Indeed, the conflict between science and fundamentalist religion has been ongoing and undoubtedly will continue. The supposition that a change has occurred over the past 400 years from a religious to a secularized society is generally agreed to be, at least in part, a result of the emergence of the modern scientific attitude.

It is useful to point out some features of the religious attitude that set it apart from the scientific attitude. I have suggested elsewhere that these two attitudes are complementary, giving rise to particular (religious) and universal (scientific) views of understanding life experience.[13] Of particular importance, the religious attitude witnesses a sacred dimension of everyday life that is unseen from the perspective of the scientific attitude.[14]

Historically, followers of the Judeo-Christian tradition experienced God directly in world events. In some instances, core "truths" of the religious attitude were revealed or became evident—for example, to the Jews at Sinai or to the Christians at Calvary. Such direct experiences, however, are no longer common. Rather, it is encounter with God that has become for many the goal of religious experience. And the revealed "truths" have become the context for the religious person's experience of everyday life. At the same time, religious groups have become fragmented because different sets of revealed "truths" have been accepted by different groups.

The scientific attitude, on the other hand, does not have at its core any revealed "truths." Although scientists accept as "truth" the possibilty of learning how the world works, what defines truth in this context is its intersubjective validation. That is, the scientific attitude aims toward a consensus view of reality.

As mentioned in Chapter 4, the conflict between the scientific and fundamentalist religious collectives has recently been concerned with such issues as creation and evolution. Science has held the advantage in this ongoing debate for one important reason: The claims of science are open to discussion. Consequently, individuals can disagree about the strength of the evidence or its interpretation, and there is the possibility that a consensus will develop among people. In contrast, the claims of fundamentalists are not open to discussion. Rather, acceptance of the claims is a prior condition for participation in the group. Moreover, contrary claims are made by different fundamentalist groups, and this situation undermines the fundamentalist position. It is difficult for an outsider (to the religious groups) to accept the possibility that all the opposing religious claims are true. Common sense suggests that this cannot be the case.

This discussion should not be interpreted to mean that people who are scientists are in any way prevented from being religious. I have known many investigators who have said that they have no use for religion. But others are as committed to religion as they are to science. Some scientists—even if they are not heavily involved with religious activities— express positive religious feelings and an interest in questions concerning the Absolute, such as the "why" of the universe or the meaning of "God." These are questions about which science has little to offer since there are no experiments that could distinguish a God-created universe from a Godless one. Indeed, science is unable to deal with the problem of whether the world could have evolved differently than it did, or whether other universes with different physical laws are possible.[15]

SCIENCE AND ETHICAL ISSUES

Scientists are sometimes accused of being amoral, of failing to take significant stands on controversial issues such as nuclear weapons or pollution. This point is usually made in the context of discussions about the use of scientific discoveries by collectives outside of science. Since scientists are willing to accept awards for research that has had beneficial consequences for society, perhaps they should also be willing to accept penalties for research that has had detrimental consequences. It is important to recognize, however, that the beneficial or detrimental uses of research are extrinsic to the scientific attitude that made the research possible in the first place. As noted earlier in this chapter, the uses to which research will subsequently be put often are unanticipated at the time that the studies were performed.

The possibility still exists that scientific societies could adopt and publicize stands on nonscientific issues. A resolution regarding the defense budget, for instance, was introduced one year at the annual meeting of the American Society of Cell Biology. Most investigators, regardless of whether they agreed or disagreed with the content of the resolution, felt that the society was not the appropriate forum for its adoption. The resolution was not related *directly* to the business of the society. The attitude expressed was that individual scientists who wanted to make strong public statements about political, social, or religious issues should do so as participants of political, social, and religious organizations. The problem with activism by scientific societies is that the members of a society are likely to belong to disparate collectives outside of science, so that group coherence is at risk when an attempt is made to adopt positions on issues of direct concern to those nonscientific collectives. Currently working in my laboratory are individuals from China, Japan, Kenya, Russia, and Taiwan, in addition to those from the United States. We share a scientific thought style much more than a political or social thought stye.

The scientific attitude itself is a potential source of ethical problems when, for instance, this attitude is transformed into a secular faith.[16] For instance, evolution has become for some people the religion of evolutionism.[17] As a theory to explain the significance of man, evolutionism diminishes human worth because it views man as an accident of nature without intrinsic value. It can be used to justify selective breeding of people and ideas of racial supremacy such as those that were at the core of Nazi philosophy. Evolutionary theory also can be applied to economics to explain success and failure in social life. According to this view, the hardships faced by those who do not succeed are a necessary part of the struggle to attain the best sort of society. In this context, poverty can be understood as an unfortunate necessity of social evolution rather than as a social evil that can be corrected.[18]

The scientific attitude as it is expressed in the modern, scientific practice of medicine also presents difficult problems. From the scientific point of view, experimental drugs and procedures need to be tested on human subjects. While animal studies provide useful preliminary information, each species has some unique physiological features that differ from humans. For instance, a yellow fever virus vaccine produced in the 1940s and safely tested on rodents and primates turned out to contain a hepatitus virus that was infectious for man. The consequence of using the vaccine without testing it on human subjects was thousands of cases of hepatitis-induced jaundice as well as many deaths.[19] Also, since humans live longer than most animal species, treatments whose adverse consequences involve a long latent period are difficult to detect in experimental animals.

Human gene therapy epitomizes the problem. On the one hand, as no good animal models exist for the enzyme deficiency diseases that are candidates for the first trials of gene therapy, human experimentation is necessary. On the other hand, current technology cannot prevent random insertion of the genes into the chromosomes of the cells that will be transplanted into the patients. Whether such random insertions will have any effect on other, normally functioning genes may not become evident until years after the clinical trials are initiated.[20]

The dilemma of human experimentation is that there are no guidelines intrinsic to the scientific attitude that can protect individuals. If people cease to be viewed as having individual human worth, then it is possible to consider them as laboratory animals. For instance, Jews in Nazi Germany were purposely infected with diseases in order to try out new cures.[21] And the problem of ethics in clinical research carried out in the United States was highlighted in a 1966 paper that described questionable scientific studies involving mentally retarded people, criminals, and military personnel.[22]

In recognition of the important ethical problems with human experimentation, limitations have been imposed on human research. The individuals who formulated and adopted these limitations were in many cases scientists, but they were acting in a social and political context. According to the current guidelines, the potential subject has the right to know precisely the nature and purpose of the research and how it might help or adversely affect him or her. Also, the individual has the right to decline to participate in the study initially or at any time after the study has begun with the assurance that it will not affect other treatment. Finally, such research must be approved in advance by university and government committees that evaluate the need and validity of the studies.

Despite all the above precautions, there remain key points regarding human experimentation that have yet to be resolved. For instance, according to some religious thought styles, individuals do not have the right or competence to decide for themselves whether to be subjects in experiments. (Similarly, according to most modern political thought styles, individuals do not have the right to commit suicide.) An opposing point of view is that individuals have an obligation to participate in human research if the research is likely to benefit others. Moreover, there is no consensus between collectives concerning who best should decide human experimentation and other ethical issues.

In the future, these problems will become even more complicated. As genetic-engineering techniques become more sophisticated, the possibility for mass-producing identical human beings will call into question the very meaning of individuality. Yet from the point of view of the scientific attitude, the feasibility of gene cloning is a sufficient reason for trying it out.

THE IMPACT OF
THE WORLD ON SCIENCE

Turning now to a consideration of how the world affects science, it is noteworthy that random events in everyday life can influence investigators and their research activities. For instance, a major snowstorm in Albany, New York, that closed down public and private transportation also stopped scientists from reaching their laboratories at the local branch of the state university. A major hurricane threatening Galveston, Texas, closed down the medical school as well as other businesses in the city.

Admittedly, these examples are extreme. At a more mundane level, one need only experience what happens when a departmental washing machine breaks down, or when the person who washes the laboratory dishes is sick: Investigators simply have less time to do experiments because they are busy cleaning glassware. More seriously, when a fire in a local circuit board shuts down the electricity, science temporarily stops and everybody goes home. But if the power were out for a long time, the long-run consequences could be disastrous because a large number of reagents used in research are stored in refrigerators, freezers, and ultra-low temperature freezers, and these reagents could deteriorate rapidly at the wrong temperature. Ultimately, the loss of some of these reagents could set back a research project by months or even longer.

Science is also dependent on proper quality control by the industries that supply equipment and supplies for research. For instance, most of the chemicals used in research laboratories are purchased, but the investigators using them rarely want to take the time to check their purity. Rather, they simply assume that the contents of the bottle match the specifications on the label. Similarly, they usually assume that analytical instruments are properly adjusted by the manufacturer. When materials or instruments used in science are different from those expected by an investigator, experiments may not work as designed and the data obtained may be misleading.

Another example of the quality-control problem concerns the water used in the laboratory. There are numerous "horror" stories about laboratories that had trouble repeating previous studies after moving from one city to another, as a consequence of differences in the mineral contents of the city water supplies.

Finally, the very availability of necessary equipment and supplies can have an impact on the direction of research. In the late 1970s, for instance, there was a shortage of fetal bovine serum. One result of this shortage was that investigators were forced to use other kinds of sera in their media for growing cells. The use of some of these other sera resulted in new patterns of cell behavior, patterns different from those previously observed with fetal bovine serum. The shortage enhanced the importance of research performed by investigators who were studying cell growth factors that could replace serum. Suddenly, their work developed major practical as well as theoretical importance, and this was accompanied by a growth spurt in companies that marketed cell growth factors.

SCIENTISTS AS PEOPLE

One sometimes gets the impression that people outside of science think that scientists are different from other people—and that the typical, mundane activities of life (eating, sleeping, etc.) are unfortunate distractions that keep the scientist from the full-time dedicated pursuit of knowledge. This view is also shared by some scientists. The implication is that one ceases to be a person and becomes a "scientist" because s/he works in a laboratory. Also implicit in this view is the notion that although scientists cannot avoid human needs, such as emotions, such limitations are a practical feature and not an inherent limitation of doing science.

For a variety of reasons, however, it would be more realistic to recognize that human limitations are an integral feature of doing science and not something added on.[23] In the first place, many of the typical activities of science also are typical of nonscience. Scientists make observations, write about their work, discuss their work with others, and read about what others have done. But there is nothing intrinsically scientific about seeing, writing, talking, and reading. And insofar as these activities are subject to human limitations, their application to science also will be limited.

In the second place, scientists, like most people, assume that the world is real and has a constant underlying order. Neither of these assumptions in itself is scientific; each is derived from the way that most people experience the world. Moreover, the distinguishing feature that permits truths to be established within science is the assumption that truth can be verified intersubjectively. As already described, intersubjectivity depends on one's view of others as individual human beings like oneself—a basic human assumption necessary for all meaningful social interactions.[24] Thus, the same interhuman recognition that is necessary for religion and politics also makes science possible.

Finally, scientists bring to their scientific work whatever assumptions they have previously accepted from other aspects of life experience (religion, politics, etc.). These assumptions may lead individual investigators to be interested in certain research problems or to view others as inappropriate. It has been suggested that the underlying purpose of such individuals as Galileo, Kepler, and Newton, who were instrumental in the development of the modern scientific thought style, was the discovery of *God's* laws. For them, God's laws were the laws of nature. Conversely, Eastern societies may have been less interested in modern science because the thought styles of these societies did not include a Creator-God who established a natural order.[25]

THE INFLUENCE OF POLITICS
ON SCIENCE

There are many different problems that in principle can be studied using accepted scientific knowledge and methodology, but the precise choice of which problems ought to be investigated is not an inherent feature of science. Therefore, collectives outside of science can direct the movement of the scientific collective as a whole. The situation is somewhat analogous to riding a bike. Owning a bike and knowing how to ride it are aspects of bike riding that are separate from choosing the direction in which to ride.

Political collectives can legislate limits on scientific research, and such proposed limitations sometimes become issues in national elections.[26] At the extreme, such legislation may be used to prevent research that might produce results contradictory to state policy. In totalitarian countries, for instance, the political collective has used its authority to regulate precisely what is done by the scientific collective. The "Lysenko affair" in genetics research in the Soviet Union is a good example of political interference in science.[27]

In 1929, T. D. Lysenko was a practical-minded agricultural specialist who believed that he had developed improved methods for seed germination and crop production. He went to work for the central government and was assigned to the plant breeding program. Eventually, he turned to politics to explain the lack of Soviet progress in this work.

> The Party and the government have set our plant-breeding science the task of creating new varieties of plants at the shortest date. . . . Nevertheless, the science of plant breeding continues to lag behind. We are convinced that the root of this evil lies in the critical state of plant biology that we inherited from methodologically bourgeois science. . . . We must fight uncompromisingly for the reconstruction of genetic plant-breeding theory . . . on the basis of the materialistic principles of development, which actually reflect the dialectics of heredity.[28]

In short, Lysenko developed the political position that Lamarckian genetics was more consistent with Marxism than Mendelian genetics. In 1936, a confrontation occurred between the pro-Lysenko and anti-Lysenko forces at a symposium entitled "Controversial Questions of Genetics and Selection." Of 46 speakers, 19 were pro-Lysenko, 17 were anti-Lysenko, and 10 were uncommitted. Four years later, the leader of academic genetics in the USSR (N. Vavilov), who had been opposed to Lysenko, was arrested and sent to prison camp. After 1948, it was illegal to teach or do research in Mendelian genetics. At the time that Lysenko lost his influence in Soviet biology in 1964, high-school textbooks contained no information about the role of the nucleus and chromosomes in heredity.

In the above example, political interests were used to impose acceptance of a particular scientific hypothesis. In U.S. politics, attention has been focused on the regulation of the kinds of research that can be performed. That is, political collectives, on the basis of their own moral judgment and in response to pressures exerted by other collectives, have often decided whether certain types of research should be carried out regardless of their feasibility. One sensitive issue has been the use of animals in research. The Animal Liberation Front—a militant organization committed to stopping the use of animals in research—has carried out raids on animal researchers at several academic institutions. In some cases, they obtained evidence of animal mistreatment that was sufficient to convince the NIH to suspend the grants that had been funding the research. In response to the concerns and demands of the animal welfare movement, new regulations for the use of animals in research have been developed. Investigators are under pressure to replace animal models with *in vitro* cell culture models, to refine methodology so as to reduce animal pain and stress, and to reduce the numbers of animals used in research.[29]

In addition to legislating restraints on scientific research, the political collective regulates science through overall budgetary allocations. In 1980, the federal budget for the NIH was more than $3 billion, split about equally between basic and applied research.[30] But the money was not simply given to the NIH. Rather, allocations for each particular institute were specified. The largest amount of money went to the institutes concerned with cancer and heart disease. Constraints were also placed on how the allocated money could be used: so much for research, so much for contracts, and so much for education. Thus, the political collective of the U.S. federal government—a group outside of science—determined in large part how biomedical science would be practiced during 1980 by specifying the amount and uses of the funds that would be spent on science. If more money had been given to the National Institute of Aging and less to the National Cancer Institute, there would have been more research on aging and less on cancer. And if more money had been given to postdoctoral training programs, more postdoctoral fellows would have been trained.

The decision regarding how much money to allocate for biomedical research depends in part on how the political collective views the health needs of the country relative to other needs. (The allocation for national defense, for instance, is almost 100 times greater than the NIH budget.) The fact that any money at all is spent on the NIH implies that the federal government recognizes the existence of health-related problems requiring a solution, and believes that NIH-funded research has the potential to solve or contribute to the solution of these problems.

As one might expect, there are some similarities, but also many differences, between the views of science policy leaders and those of members of the public knowledgeable about science with regard to research funding priorities (see Table 6.1).[31] Assuming a no-growth budget, the major area in which the public preferred to see funding increased was disease-specific medical research (82%). This area was of much lower priority (34%) for science policy leaders.

Since the federal government represents the public, the allocation of funds within the NIH budget in part reflects public health needs. Recently, the greatest allocation of funds has been for research on cancer and heart disease, both of which, as the major causes of death in the United States, are greatly feared. When acquired immunodeficiency syndrome (AIDS) became a general public concern, money was specifically allocated for studies in this area. In the early phase of AIDS research, however, significant questions were raised about the amount of money made available and the way in which the money was spent.[32] It has been suggested that the U.S. government's initial response would have been different if the high-risk groups in the United States (i.e., gay men and drug users) had been less controversial.

Limitations on the level of funding for the NIH depend on the federal government's assessment of the value of biomedical research relative to that of other government-sponsored programs. Most groups seeking federal funds for their activities have lobbyists who make known to the political collective as well as to the public the reasons for which their groups' activities should be supported. When lobbying efforts are successful, as in defense-related programs, funds are maintained at a high level even when those for other programs are decreased. When lobbying efforts are unsuccessful, as has recently been the case for many social programs including scientific research, allocations may be cut or eliminated altogether.

Scientific collectives engage in lobbying efforts on behalf of science in a variety of ways. Organizations such as the National Academy of Sciences have committees to study and make recommendations—often to the federal government—on policies relating to science. Similarly, many scientific organizations have public policy committees that make recommendations to the organization's membership regarding important issues that should be supported or opposed. Recognizing the importance of getting members of Congress to understand more about science, some professional societies have sponsored congressional fellow programs in which a scientist who is a member of the society is funded for a year's work in a congressional office.

Scientific collectives also make the current status of scientific research known directly to the public. Universities and professional societies distribute press releases regarding work anticipated to be of general interest. Scientific prizes are awarded to publicize very important find-

TABLE 6.1
Areas of Science to Receive Increased Funding in a No-growth Budget

Group Surveyed	Designated Area of Research	% of Individuals in Group Favoring an Increase
Public interested in science	Disease-specific medical research	82
	Human learning processes	67
	Science and engineering education	57
	Basic chemistry	47
	Space exploration	47
	Engineering	46
	Basic physics	46
	Basic biological research	44
	Behavior in complex organizations	39
	Weapons research and development	36
	Mathematics	32
Science policy leaders	Science and engineering education	55
	Basic biological research	54
	Human learning processes	42
	Basic chemistry	41
	Engineering	39
	Basic physics	37
	Disease-specific medical research	34
	Space exploration	34
	Behavior in complex organizations	33
	Mathematics	22
	Weapons research and development	11

Adapted from: National Science Board, 1983, "Science Indicators 1982," Government Printing Office, Washington, D.C.

TABLE 6.2
Public Support for Increased Spending on Different Problems

Problem Area	Percentage of Public Favoring Increased Support	
	Public Has Sustained Interest in Science	Public Lacks Sustained Interest in Science
Reducing the crime rate	71	75
Improving education	63	61
Reducing and controlling pollution	63	47
Improving health care	59	62
Conducting scientific research	49	25
Helping low-income persons	42	45
Exploring space	39	10
Developing and improving national defense weapons	32	31

Adapted from: National Science Board, 1983, "Science Indicators 1982," Government Printing Office, Washington, D.C.

ings. Some professional societies—for instance, the American Association for the Advancement of Science—even invite the participation of nonscientists as well as scientists at their regular meetings. As a consequence, these nonscientists can learn about the current status of different scientific fields and get acquainted with individual scientists.

Despite the efforts noted above, however, the lobbying activities conducted by the scientific collective are not as extensive as those that occur on behalf of other collectives. For instance, there are no pro-science political action committees. One member of Congress warned a group of scientists that biomedical researchers should become more involved with Congress, not only with regard to allocations but also to help the political collective understand how the scientific collective functions.[33] Most scientists seem to believe that the benefits of biomedical research as well as other aspects of science are self-evident. Public polls, however, show that compared to other societal needs, support for conducting scientific research and exploring space is only of moderate priority even for members of the public who take an active, continuing interest in science. Moreover, such increases are of low priority for members of the public who have no sustained interest in science (Table 6.2).[34]

THE INFLUENCE OF RELIGION
ON SCIENCE

Unlike political collectives, religious collectives have no direct control over the scientific domain—at least not in those countries where there is a separation of church and state. The influence of religious collectives, therefore, has to be applied either through individual scientists or through political collectives. Because fundamentalist religious collectives view the theoretical aspects of science as undermining religion, they have attempted to prevent public schools from teaching the scientific viewpoint (e.g., theories of the earth's origin and of evolution of life). These groups would like to have scientific ideas about such subjects omitted from the school curriculum altogether or taught as very tentative theories (in the latter case they want Creation Science taught as an opposing and equally likely theory).

When laws requiring the teaching of Creation Science have been adopted by state legislatures in response to pressures exerted by the religious collective, the legality of the laws has been challenged in the courts. Since the Constitution requires that religion not be taught in the public schools, attempts to teach religion under the guise of science generally have been ruled unconstitutional.[35] Nevertheless, the efforts by fundamentalist groups to dilute or delete from school curricula what science has learned about the origin of the earth and the evolution of life have had some success. Contributing to their success are the science textbooks published by companies that decrease the presentation of information that the publishers anticipate will reduce adoptions of their books by local school systems. The result has been production of less sophisticated texts, which contribute to the lack of rigor in public science education.

Beyond the theoretical issues, religious collectives become involved in the practical applications of science. An important example is the opposition of some religious groups to abortion. Science has made it possible to determine whether a fetus is abnormal by using physical techniques such as sonography or genetic-screening methods. But these techniques can show only that a particular fetus is unlike "typical" fetuses. Whether or not an "atypical" fetus should be aborted is an issue outside of science. It is an issue that concerns individual human rights, and one in which religious collectives have a considerable interest.[36]

SCIENCE AND THE PUBLIC

It should be evident from the preceding discussion that individuals engaged in political and religious activities can have a considerable influence on the activities of scientists. It is important, therefore, that these nonscientists understand what doing science means. Public knowledge about science, however, is quite limited. A 1981 study[37] found that *only about 20% of the public have an active, ongoing interest in science* and keep informed. Another 20% identified themselves as very interested in science but not well informed. Finally, the remaining 60% of the public recognize the importance of science but are not interested. Those individuals who are interested learn about science from formal science education, from the technological applications of science, from the public media, and from interactions with individual scientists.

As presented at the level of science education, the image of science tends to be idealistic. The theoretical and technical contributions of science are usually discussed without clarification as to how they were accomplished. Students have learned that there is a "scientific method," and scientists generally have made little effort to correct this misrepresentation. Consequently, few people understand that there are *many* scientific methods, and that science works as a collective, tentative approach to understanding reality, limited in its capacity to shed light on important aspects of life experience. Unfortunately, many people have the impression that science is an enterprise carried out by individuals, which before long will be able to explain all of reality, and that appropriate applications of scientific methodology can solve any and all of the problems facing society.

Beyond education, individuals encounter science primarily through its applications. The relationship between basic research and its subsequent applications often is not appreciated. The contributions of basic research that lead to development of technologies unrelated to the original research, such as the use of enzymes by the food and clothing industries, are not known. But when applications of science are seen as dangerous to society (e.g., as a result of environmental pollution or potential nuclear disasters), the tendency is to hold science responsible. On the other hand, the potentially de-individualizing aspect of science— the treatment of individuals as typical members of a class—is not recognized as a problem.

The public view of science is also learned from the media, including magazines, newspapers, television, and films. A study presented at the 1985 annual meeting of the American Association for the Advancement of Science reported that scientists on television have a positive image.[38] They are presented as rational beings, if a bit older and stranger than other professionals. For every villainous scientist there are five virtuous ones. Occasionally, science and innovation are the subject matter of programs such as the PBS series "Quests for the Killers" based on June Goodfield's book.[39] For the most part, however, the goal of the television programs in which scientists are portrayed is no more than mere entertainment.

As might be expected, the destructive consequences of science (e.g., nuclear disasters, environmental pollution) are major items in the news media. Sometimes, however, news articles that could focus on science from a positive perspective present the opposite point of view. When Barbara McClintock was awarded the Nobel Prize, most of the news stories were about her struggle and how science had failed to recognize her accomplishments. They made little attempt to understand why her accomplishments were not fully appreciated earlier or how progress in the field made possible her subsequent recognition. In general, the presence of conflicting views within science is presented as something unusual[40] rather than as an essential and typical feature of science.

A science article in the July 10th (1984) issue of the *Jerusalem Post* stated that an investigator from the Weizmann Institute had recently been awarded the FEBS prize at the meeting of the Federation of European Biochemical Societies held in Moscow, where there were 90 Israeli scientists attending among 5,000 delegates. The article, entitled "Weizmann scientist honoured in Moscow," was on page 3 in a vertical column that contained three other articles. The articles could have been typeset in any arrangement, but the actual order by headlines was "19 drowning victims ignored safety rules" at the top of the page; then "Girl alleges she was raped during boat ride"; then "Weizmann scientist . . ."; and finally, "Man found dead in police lock-up." This order suggests how important the science article was viewed by the editor relative to the other stories. The extraordinary fact that science constitutes an international collective crossing geographic boundaries between countries that do not have diplomatic relations with each other (e.g., Israel and Russia) was not mentioned. One imagines that if this international scientific meeting had been a sporting event, the level of coverage would have been much different.

Direct interactions between scientists and nonscientists are another way in which nonscientists learn about science. Unfortunately, many scientists have difficulty explaining how science works; the language of science is so specialized that it is difficult to explain to the nonscientist even what questions are under study. In the *Jerusalem Post* article mentioned above, the award was for "contribution to the understanding of cell movement and the means by which the activity of the cell membrane is controlled." It would be very difficult for anyone not already familiar with the scientist's work to understand much from this description of the research.

On balance, the public seems to be positive but cautious about science. The general belief is that science may solve society's problems

TABLE 6.3
Public Views About Science and Technology

Statement Posed	% Responses		
	Agree	Disagree	Don't Know
1. Through science and technology we can continue to raise our standard of living.	80	18	3
2. Most problems can be solved by applying more and better technology.	77	21	2
3. Scientists can solve any problem we might face if they are given enough time and money.	42	55	3
4. Science and technology do as much harm as good.	56	39	4
5. Science and technology often get out of hand, threatening society instead of saving it.	77	21	2

Adapted from: National Science Board, 1983, "Science Indicators, 1982," Government Printing Office, Washington, D.C.

and improve our standard of living, but some people view science as a potential threat (Table 6.3).[41] Comparing the institutions about whose leadership the public say they have a "great deal of confidence," we find that medicine (46%) and the scientific community (38%) are at the top of the list, slightly higher than organized religion (32%) and much higher than the Congress (13%).[42] Nevertheless, there is the constant possibility that science will fail to meet the public's expectations, in which case the view can develop that scientists are spending their time studying irrelevant problems and wasting public funds. Thus, it is essential to the continued survival of the scientific collective that others outside of science be encouraged to understand how science works, what it can accomplish, and the ways in which it is limited.

Concluding Comments

7 I have tried in this book to present a comprehensive overview of science. As an activity, science can be considered at three, interdependent levels. At the first level is the individual investigator who makes observations, formulates hypotheses, and so forth. At the second level is the scientific collective that first recruits and then instructs the individual investigator in the appropriate ways of doing science and subsequently keeps the investigator's work under constant evaluation. At the third level, finally, is daily life, which provides the subject matter for scientific investigation and continuously influences and is influenced by science.

Critical to understanding the meaning of science is recognition of the scientific attitude as a particular way of thinking about and understanding the reality of the world. Application of the scientific attitude constitutes a scientific domain within experience, but this domain is neither comprehensive nor absolute. That is, there are aspects of experience that cannot be understood by means of the scientific attitude, and conclusions reached through the scientific attitude are necessarily tentative.

Individual investigators do not systematically formulate hypotheses based on their observations. Rather, hypotheses begin as intuitive guesses, often based on very little evidence. Moreover, observations and hypotheses are made in the context of previously accepted observations and hypotheses that are taken for granted by the investigator (i.e., in the context of the investigator's thought style). These prior assumptions determine and limit what can be seen or thought by the investigator. Seeing or thinking of something new, therefore, means that it can be placed within the context of the preexisting beliefs that make up the thought style. Assimilation of the new information by the thought style introduces changes (in other words, the thought style accommodates), and the modified thought style can then be used for further acquisition of additional information.

Since the scientific thought style constrains the ability of investigators to make observations and conceptualize hypotheses, expansion of the thought style is beneficial to the movement of science. Such expansion can be promoted when individual investigators move beyond their particular fields to apply their thought styles to new problems, or when groups of investigators with different thought styles collaborate on the same problem.

Controversies between investigators can be understood in a general sense as clashes between thought styles rather than in a personal sense as one investigator being right and the other wrong. The key to understanding what underlies a controversy is the difference in the starting assumptions of the investigators involved. The status of the individuals in the scientific thought collective (student versus advisor or new investigator versus established investigator) is irrelevant. It is possible to learn a great deal by analyzing the differences in starting assumptions, but little is learned when status in the scientific collective is used to settle a dispute.

In the context of doing particular experiments, it is essential that the investigator recognize the implicit starting assumptions that make the experiments possible in the first place. Every experiment should be understood as testing both explicit and implicit hypotheses. Otherwise, results that call into question the implicit hypotheses often will be overlooked, and an opportunity to learn new and important information will be missed. The major influence of luck in science is that investigators sometimes get unexpected opportunities to see things in new ways.

Biological hypotheses cannot avoid functional considerations. As a science of life, biology ultimately focuses on interacting systems. A reductionistic approach can elucidate the individual characteristics of a system, but a holistic approach is required to understand physiological interrelationships.

What distinguishes the scientific attitude from other attitudes is the assumption of universal, intersubjective validity. Truth in science means attaining a consensus view of reality. If this goal is to be reached, it must be possible, in principle, for an investigator's observations and hypotheses to be studied and verified intersubjectively. Therefore, individual investigators do science with the expectation that other investigators will be able to make the same observations and reach the same conclusions. A corollary of this expectation is that those aspects of the world that are not believed by scientists to be subject to intersubjective verification are excluded from the scientific domain. The requirement for intersubjective verification is fulfilled by a network of investigators (the scientific collective) who can adopt the scientific attitude and evaluate each other's work.

Most people learn about the "scientific method" rather than about the scientific attitude. While the "scientific method" is an ideal construct, the scientific attitude is a way people have of looking at the world. Doing science includes many methods; what makes them scientific is their acceptance by the scientific collective. This acceptance depends on assumptions regarding already accepted knowledge as well as assumptions about how new information can be gathered appropriately (i.e., the collective's prevailing thought style). The thought style, however, is pragmatic and accepts procedures only so long as they appear to be useful. Thus, while the intersubjective feature of science persists, the details of how one does science change with the times.

It is important that the individual investigator understands how the scientific collective operates. Convincing oneself of new scientific findings (heuristic research) and convincing others (demonstrative research) are separate activities. Investigators who have inspired ideas but cannot communicate them effectively to the scientific collective will have difficulty publishing their work and obtaining grant support. The pragmatics of the scientific enterprise dictate that learning how to communicate with the collective is just as important as learning something new to communicate. Unfortunately, students are often not taught how to communicate effectively as part of doing science.

The funding system is a weak link in the scientific network of the United States. The mechanism of evaluation of grant applications discourages novelty and innovation. Desirable changes would include longer funding periods, decreased emphasis on specific aims and experiments planned, and increased emphasis on rationale and overall approach. The method of assigning relative rankings to grants and then funding them on an all-or-none basis is arbitrary and self-defeating. Investigators who are doing important research and getting very good priority scores are sometimes forced to shut down or curtail operations unnecessarily. Since grants are approved on a relative basis, they could be funded on a relative basis.

Compounding the problem of evaluating grants has been the overall lack of sufficient funding for research. The scientific collective has been only partially successful in convincing those outside science of the need for increased basic research support. The general lack of public interest and understanding of science is indeed cause for concern.

The truths established according to the scientific attitude, though limited and tentative, have had an extraordinary impact on the world. The observations and hypotheses made by scientists in one year have the potential to change the way people live and die in the next. People hope that science will solve their problems, but they should also recognize that the scientific attitude, when it is directed toward people, ignores the individual and is concerned only with what is typical. Private experiences that cannot be shared intersubjectively are discounted. Consequently, the scientific attitude, when applied outside of science, can act as a de-individualizing force.

Recognition of this de-individualizing feature of science ought to lead to an expansion of the interactions between scientists and members of other collectives. Scientists need to understand better the views of the religious and political collectives, and nonscientists need to learn how science works. Ethical problems with biomedical research constitute a problem for everybody, scientists and physicians, politicians and clergy. As it now stands, however, scientists rarely study biomedical ethics, and in some medical schools this subject is peripheral rather than central. Changes in the medical and graduate school curricula could easily remedy thse oversights.

The direction and scope of science are determined in part by collectives outside of science. If other collectives adopt the view that science is no longer useful, or that science is threatening, then traditional support for scientific research will decline. On the other hand, if the limitations of science are recognized and scientists can convey how the scientific attitude works, then it should be possible for people to develop a more realistic expectation of what science has to offer.

Overall, science presents a paradox. Its practical effects indicate that science is more than mere belief, but the truths of science change with the thought style. An awareness of this paradox can act as an antidote to scientific hubris and prevent scientists from becoming dogmatic. Nevertheless, the goal of the scientific attitude is to learn truths that have universal, intersubjective validity. For the moment, therefore, this attitude is unique in its contributions to the study and understanding of those aspects of reality shared by all people.

Notes and
References

This reference list is neither inclusive nor representative. The works cited are simply the sources from which I first learned about the ideas described in this book.

I wish to draw special attention to the following four sources because of their impact on me. One of the first philosophy books with which I became acquainted was William James's "Some Problems in Philosophy" (Greenwood Press, 1968, originally published in 1911). James wrote about philosophical issues in a way that I, as a scientist, could understand clearly. This was no mean feat given that the jargon of philosophy is exceedingly difficult for those who have no formal background in philosophy studies. It was through James's book that I became aware of the important differences between perceptual experiences and conceptual ideas. In my subsequent reading, I have never found another philosopher able to describe complex ideas with such clarity.

David Hull's "Philosophy of Biological Science" (Prentice Hall, 1974) was very helpful as an introduction to the philosophy of science. I was familiar with the examples he used, so I was able to attend to the conceptual issues he raised. Much of modern philosophy of science uses examples from quantum physics, and it was difficult for me to understand the philosophy while struggling to understand the science.

The ideas of "thought collectives" and "thought styles" are borrowed from Ludwik Fleck's "Genesis and Development of a Scientific Fact" (University of Chicago Press, 1979, originally published in German in 1935). I have found few people who are familiar with this book—and I cannot remember who first told me to read it—but as Thomas Kuhn says in his Foreword to the translated edition, it is a brilliant study. Fleck's book is about the development of scientific ideas as well as about the social relations among scientists. He weaves these themes around a history of syphilis and the development of modern immunology. His book was the first place I found a description of everyday science that fit with my experience.

The last book I want to mention is Alfred Schutz's "The Phenomenology of the Social World" (Northwestern University Press, 1967, originally published in German in 1932). Unlike the other works mentioned above, this book is not introductory. I would recommend it only to those readers acquainted with the jargon and central themes of phenomenology. I was fortunate to study the book with Maurice Natanson, who had been Schutz's student. This work was the source for my understanding of the idea of intersubjectivity and the importance of intersubjectivity in social relationships. Moreover, it was this book that led me to think about the features typical of doing science, which is what my book is about.

NOTES TO PREFACE

1. Medawar, P.B., 1984, "The Limits of Science," Harper and Row, New York.

NOTES TO CHAPTER 1

1. Adapted from Watson, J.D., 1968, "The Double Helix," Atheneum Publishers, New York.
2. Thomas, L., 1974, "The Lives of a Cell: Notes of a Biology Watcher," Viking Press, New York.
3. Klemke, E.D., Hollinger, R., and Kline, A.D., 1980, "Introductory Readings in the Philosophy of Science," Prometheus Books, New York.
4. Ziman, J., 1980, What is science. In "Introductory Readings in the Philosophy of Science," ed. E.D. Klemke, R. Hollinger, and A.D. Kline, Prometheus Books, New York, pp. 35–54.
5. Kockelmans, J.J., 1970, The mathematization of Nature in Husserl's last publication. In "Phenomenology and the Natural Sciences," ed. J.J. Kockelmans and T.J. Kisiel, Northwestern Univ. Press, Evanston, pp. 45–67. See also Heelan, P.A., 1983, "Space-Perception and the Philosophy of Science," Univ. of California Press, Berkeley.
6. Rosak, T., 1969, "The Making of a Counterculture," Doubleday, Garden City, N.Y.
7. Minkowski, E., 1970, Prose and poetry. In "Phenomenology and the Natural Sciences," ed. J.J. Kockelmans and T.J. Kisiel, Northwestern Univ. Press, Evanston, pp. 239–250.
8. Zaner, R.M., 1970, "The Way of Phenomenology," Western Publishing Co., New York.

NOTES TO CHAPTER 2

1. Pryde, D., 1971, "Nunaga: Ten Years of Eskimo Life," Walker and Co., New York.
2. Hardin, D., 1957, The threat of clarity. American Journal of Psychiatry 114: 392–396.
3. Dyson, R.D., 1974, "Cell Biology: A Molecular Approach," Allyn and Bacon, Boston.

4. "Random House Dictionary of the English Language," 1967, Random House, New York.

5. James, W., 1968, "Some Problems in Philosophy," Greenwood Press, New York.

6. Kisiel, T.J., 1970, Merleau-Ponty on philosophy and science. In "Phenomenology and the Natural Sciences," ed. J.J. Kockelmans and T.J. Kisiel, Northwestern Univ. Press, Evanston, pp. 251–311.

7. Gibson, E.J., 1969, "Principles of Perceptual Learning and Development," Prentice-Hall, Englewood Cliffs, N.J.

8. Gurwitsch, A., 1964, "The Field of Consciousness," Univ. of Pittsburgh Press, Pittsburgh.

9. Heelan, P.A., 1983, "Space-Perception and the Philosophy of Science," Univ. of California Press, Berkeley.

10. Wadsworth, B.J., 1971, "Piaget's Theory of Cognitive Development," David McKay Company, Inc., New York.

11. Polanyi, M., 1967, "The Tacit Dimension," Doubleday, Garden City, N.Y.

12. Gurwitsch, "The Field of Consciousness."

13. James, "Some Problems in Philosophy."

14. Wadsworth, "Piaget's Theory."

15. Gurwitsch, A., 1966, Some aspects and development of Gestalt psychology. In "Studies in Phenomenology and Psychology," ed. J. Wild and J.M. Edie, Northwestern Univ. Press, Evanston, pp. 3–55. See also Hanson, N.R., 1980, Observation. In "Introductory Readings in the Philosophy of Science," ed. E.D. Klemke, R. Hollinger, and A.D. Kline, Prometheus Books, New York, pp. 152–163.

16. Fleck, L., 1979, "Genesis and Development of a Scientific Fact," Univ. of Chicago Press, Chicago.

17. Goodfield, J., 1985, "Quest for the Killers," Birkauser Boston Inc., Cambridge, Mass.

18. Fleck, "Genesis and Development"; Kuhn, T.S., 1962, "The Structure of Scientific Revolutions," Univ. of Chicago Press, Chicago.

19. Bechtel, W., 1984, The evolution of our understanding of the cell: A study in the dynamics of scientific progress. Studies in the History and Philosophy of Science 15: 309–356.

20. Goodfield, "Quest for the Killers."

21. Davis, B.D., 1980 Frontiers of biological science. Science 209: 78–89.

22. Watson, J.D., 1968, "The Double Helix," Atheneum Publishers, New York.

23. Gurwitsch, "The Field of Consciousness."

24. Slatkin, M., 1985, The descent of genes. Science 85: 80–81.

25. Hall, S., 1985, The fate of the egg. Science 85: 40–48.

26. Weiss, P.A., 1969, The living system: Determinism stratified. In "Beyond Reductionism," ed. A. Koestler and J. R. Smythies, Macmillan, New York, pp. 3–55. See also Polanyi, M., 1968, Life's irreducible structure. Science 162: 1308–1312.

27. Glasstone, S., 1947, "Thermodynamics for Chemists," D. Van Nordstrand Co., Princeton.

28. Harvard Ad Hoc Committee, 1968, A definition of irreversible coma. Journal of the American Medical Association 205: 337–340.

29. Schutz, A., 1967, "The Phenomenology of the Social World," tr. G. Walsh and F. Lehnert, Northwestern Univ. Press, Evanston.

NOTES TO CHAPTER 3

1. Schutz, A., 1967, "The Phenomenology of the Social World," tr. G. Walsh and F. Lehnert, Northwestern Univ. Press, Evanston.

2. W. Whewell, quoted in Medawar, P.B., 1984, "The Limits of Science," Harper and Row, New York.

3. Ross, R., Glomset, J., Kariya, B., and Harker, L., 1974, A platelet-dependent serum factor that stimulates the proliferation of arterial smooth muscle cells in vitro. Proceedings of the National Academy of Science USA 71: 1207–1210.

4. Hull, D., 1974, "Philosophy of Biological Science," Prentice-Hall, Englewood Cliffs, N.J.

5. Gurwitsch, A., 1964, "The Field of Consciousness," Univ. of Pittsburgh Press, Pittsburgh.

6. Goodfield, J., 1985, "Quest for the Killers," Birkauser Boston Inc., Cambridge, Mass.

7. Wilson, D., 1976, "In Search of Penicillin," Alfred A. Knopf Publishers, New York.

8. Ibid.

9. Ibid.

10. Humphrey, J.H., 1984, Serendipity in immunology. Annual Review of Immunology 2: 1–21.

11. Medawar, "The Limits of Science."

12. Goodfield, "Quest for the Killers."

13. Medawar, "The Limits of Science."

14. Eddington, A.S., 1939, "Philosophy of Physical Science," Cambridge University Press, Cambridge.

15. Medawar, "The Limits of Science."

16. Goodfield, "Quest for the Killers."

17. Polanyi, M., 1969, "Knowing and Being," ed. M. Greene, Univ. Chicago Press, Chicago; Medawar, P.B., 1969, "Induction and Intuition in Scientific Thought," Methuen Press, London.

18. Jaffe, B., 1976, "Crucibles: The Story of Chemistry from Ancient Alchemy to Nuclear Fusion," Dover Publications, Mineola, N.Y.

19. Hull, "Philosophy of Biological Science."

20. Nickles, T., 1980, "Scientific Discovery, Logic, and Rationality," D. Reidel Publishing Co., Boston.

21. Kuhn, T.S., 1962, "The Structure of Scientific Revolutions," Univ. of Chicago Press, Chicago.

22. Popper, K., 1980, Science: Conjectures and refutations. In "Introductory Readings in the Philosophy of Science," ed. E.D. Klemke, R. Hollinger, and A.D. Kline, Prometheus Books, New York, pp. 19–34.

23. Compare with Thagard, P.R., 1980, Why astrology is pseudoscience. In "Introductory Readings in Philosophy of Science," ed. E.D. Klemke, R. Hollinger, and A.D. Kline, Prometheus Books, New York, pp. 66–75.

24. Rudner, R., 1980, The scientist qua scientist makes value judgements. In "Introductory Readings in Philosophy of Science," ed. E.D. Klemke, R. Hollinger, and A.D. Kline, Prometheus Books, New York, pp. 231–237.

25. Mowry, B., 1985, From Galen's theory to William Harvey's theory: A case study in the rationality of scientific theory change. Studies in the History and Philosophy of Science 16: 49–82.

26. Burtt, E.A., 1954, "The Metaphysical Foundations of Modern Science," Doubleday, Garden City, N.Y.

27. Hull, "Philosophy of Biological Science."

28. Compare with Eldridge, N., 1985, "Unfinished Synthesis: Biological Hierarchies and Modern Evolutionary Thought," Oxford Univ. Press, New York.

29. Polanyi, M., 1968, Life's irreducible structure. Science 162: 1308–1312.

NOTES TO CHAPTER 4

1. Medawar, P.B., 1984, "The Limits of Science," Harper and Row, New York.

2. Compare with Laudan, L., 1980, Why was the logic of discovery abandoned? In "Scientific Discovery, Logic, and Rationality," ed. T. Nickles, Reidel Publishing Co., Boston, pp. 173–183.

3. Kuhn, T.S., 1962, "The Structure of Scientific Revolutions," Univ. of Chicago Press, Chicago.

4. Fleck, L., 1979, "Genesis and Development of a Scientific Fact," Univ. of Chicago Press, Chicago.

5. Jukes, T.H., 1983, The creationist attack on science. Introduction: The magnitude of the threat. Federation Proceedings 42: 3022–3024.

6. Ibid.

7. Nobel Foundation, 1972, "Nobel Lectures in Physiology or Medicine, 1963–1970," Elsevier Publishing Co., Amsterdam.

8. Astrin, S.M., and Rotherg, P.G., 1983, Oncogenes and cancer. Cancer Investigation 1: 355–364.

9. Northrop, J.H., 1961, Biochemists, biologists, and William of Occam. Annual Review of Biochemistry 30: 1–10.

10. Darwin, C., 1966, "On the Origin of Species," Harvard Univ. Press, Cambridge, Mass.

11. Hull, D.L., Tessner, P.D., and Diamond, A.M., 1978, Planck's principle. Science 202: 713–722.

12. Gilbert, G.N., and Mulkay, M., 1984, "Opening Pandora's Box," Cambridge Univ. Press, Cambridge.

13. Mitchell, P., 1976, Vectorial chemistry and the molecular mechanisms of chemiosmotic coupling: Power transmission by proticity. Biochemical Society Transactions 4: 412–429.

14. Gilbert and Mulkay, "Opening Pandora's Box."

15. Mitchell, Vectorial chemistry.

16. West, E.S., Todd, W.R., Mason, H.S., and Van Bruggen, J.T., 1966, "Textbook of Biochemistry," Macmillan, New York.

17. Lehninger, A.L., 1975, "Biochemistry," Worth Publishers, New York.

18. Fleck, "Genesis and Development."

19. Ibid.

20. Mowry, B., 1985, From Galen's theory to William Harvey's theory: A case study in the rationality of scientific theory change. Studies in the History and Philosophy of Science 16: 49–82.

21. W. Tuckwell, quoted in Hughes, A., 1959, "A History of Cytology," Abelard-Schuman Publishers, New York.

22. Kuhn, "The Structure of Scientific Revolutions."

23. Physicians for the twenty-first century: The GPEP report, 1984, Association of American Medical Colleges, Washington, D.C.

24. Culliton, B.J., 1984, Medical education under fire. Science 226: 419–420.

25. National Academy of Sciences, 1982, "An Assessment of Research Doctorate Programs in the United States—Biological Science," National Academy Press, Washington, D.C.

26. Best places to be for a Ph.D., 1983, Changing Times 37: 64–67.

27. Buber, M., 1948, "Tales of the Hasidim: The Later Masters," Schocken Books, New York.

28. Zuckerman, H., 1977, "Scientific Elite: Nobel Laureates in the United States," Macmillan, New York.

29. Ziman, J., 1980, What is science? In "Introductory Readings in the Philosophy of Science," ed. E.D. Klemke, R. Hollinger, and A.D. Kline, Prometheus Books, New York, pp. 35–54.

30. National Research Council, 1981, "Postdoctoral Appointments and Disappointments," National Academy Press, Washington, D.C.

31. Ibid.

NOTES TO CHAPTER 5

1. Anderson, R.C., Narin F., and McAllister, P., 1978, Publication ratings versus peer ratings of universities. Journal of the American Society of Information Science (March): 91–103.

2. Zuckerman, H., 1977, "Scientific Elite: Nobel Laureates in the United States," Macmillan, New York.

3. National Science Board, 1983, "Science Indicators 1982," Government Printing Office, Washington, D.C.

4. de Solla Price, D.J., 1966, The science of scientists. Medical Opinion and Review (July): 88–97.

5. Compare with Abelson, P.H., 1980, Scientific communication. Science 209: 60–62.

6. Fleck, L., 1979, "Genesis and Development of a Scientific Fact," Univ. of Chicago Press, Chicago.

7. Heelan, P.A., 1983, "Space-Perception and the Philosophy of Science," Univ. of California Press, Berkeley.

8. Gilbert, G.N., and Mulkay, M., 1984, "Opening Pandora's Box," Cambridge Univ. Press, Cambridge.

9. Huth, E.J., 1982, "How to Write and Publish in the Medical Sciences," ISI Press, Philadelphia, Pa.

10. Broad, W., and Wade, N., 1982, "Betrayers of the Truth," Simon and Schuster, New York.

11. Hull, D., 1985, Openness and secrecy in science: Their origins and limitations. Science, Technology, and Human Values 51: 4–13.

12. Abelson, Scientific communication.

13. Barber, B., 1961, Resistance by scientists to scientific discovery. Science 134: 596–602.

14. Hull, D.L., 1985, Bias and commitment in science: Phenetics and cladistics. Annals of Science 42: 319–338.

15. Ibid.

16. Goffman, W., and Newill, V.A., 1964, Generalization of epidemic theory. Nature 204: 225–228.

17. Journal Citation Reports, 1982, ed. E. Garfield, ISI Press, Philadelphia, Pa.

18. Garfield, E., 1983, "Citation Indexing," ISI Press, Philadelphia, Pa.

19. Ibid.

20. Cole, S., 1970, Professional standing and the reception of scientific discoveries. American Journal of Sociology 76: 286–306; Stent, G.S., 1972, Prematurity and uniqueness in scientific discovery. Scientific American 227: 84–93.

21. Garfield, "Citation Indexing."

22. Small, H.G., 1973, Co-citation in the scientific literature: A new measure of the relationship between two documents. Journal of the American Society of Information Science 24: 265–269.

23. Small, H.G., 1977, A co-citation model of a scientific speciality and a longitudinal study of collagen research. Social Studies of Science 7: 139–166.

24. Journal Citation Reports, 1982.

25. Ibid.

26. Fleck, "Genesis and Development."

27. Culliton, B.J., 1984, Fine-tuning peer review. Science 226: 1401.

28. Wyngaarden, J.B., 1984, Nurturing the scientific enterprise. Science 223: 361–364.

29. McLoughlin, W.J., 1983, Stabilization of biomedical sciences: An achievable priority. Cancer Investigations 1: 351–354.

30. Why grant proposals are unsuccessful. N.I.H. Health Grants and Contracts Weekly (March 31, 1981), Bethesda, Md.

31. Braber, D.W., 1985, Innovation and academic research. Nature 316: 401–402.

32. American Society of Biological Chemists, 1986, Serial 6-1087, Bethesda, Md.

33. Garfield, E., 1977, The 259 most-cited primary authors, 1961–1975. II. The correlation between citedness, Nobel Prizes, and academy memberships. Current Contents 50: 5–15.

34. Compare with Fleck, "Genesis and Development," and Faberge, A.C., 1982, Open information and secrecy in science. Perspectives in Biology and Medicine 25: 263–278.

35. Nelkin, D., 1982, Intellectual property: The control of scientific information. Science 216: 704–708.

36. Broad and Wade, "Betrayers of the Truth."

37. Babbage, C., 1969, "Reflections on the Decline of Science in England and on Some of Its Causes," Gregg International, London.

38. Smith, R.J., 1985, Scientific fraud probed at AAAS meeting. Science 228: 1292–1294.

39. Broad and Wade, "Betrayers of the Truth."

40. Ibid.

41. Ibid.

NOTES TO CHAPTER 6

1. Radner, D., and Radner, M., 1982, "Science and Unreason," Wadsworth Publishing Co., Belmont, Ca.

2. Maxwell, G., 1980, The ontological status of theoretical entities. In "Introductory Readings in the Philosophy of Science," ed. E.D. Klemke, R. Hollinger, and A.D. Kline, Prometheus Books, New York, pp. 175–184.

3. Heelan, P.A., 1983, "Space-Perception and the Philosophy of Science," Univ. of California Press, Berkeley.

4. Polanyi, M., 1969, "Knowing and Being," ed. M. Greene, Univ. Chicago Press, Chicago.

5. Hull, D., 1974, "Philosophy of Biological Science," Prentice-Hall, Englewood Cliffs, N.J.

6. Carmell, A., 1972, Freedom, providence, and the scientific outlook. In "Challenge," ed. A. Carmell and C. Domb, Feldheim Publishers, New York.

7. Jones, R.S., 1982, "Physics as Metaphor," Univ. of Minnesota Press, Minneapolis.

8. Shibayama, Z., 1974, "Zen Comments on the Mumonkan," tr. S. Kudo, Harper and Row, New York.

9. Ibid.

10. James, W., 1961, "The Varieties of Religious Experience," Macmillan, New York.

11. Eisenberg, L., 1983, The subjective in medicine. Perspectives in Biology and Medicine 27: 48–61.

12. Garfield, E., 1981, The economic impact of research and development. Current Contents 51: 337–347.

13. Grinnell, F., 1986, Complementarity: An approach to understanding the relationship between science and religion. Perspectives in Biology and Medicine 29: 292–301.

14. Durkheim, E., 1965, "The Elementary Forms of Religious Life," Free Press, New York.

15. Hospers, J., 1980, What is explanation? In "Introductory Readings in the Philosophy of Science," ed. E.D. Klemke, R. Hollinger, and A.D. Kline, Prometheus Books, New York, pp. 87–103.

16. Feyerabend, P., 1980, How to defend science against society. In "Introductory Readings in the Philosophy of Science," ed. E.D. Klemke, R. Hollinger, and A.D. Kline, Prometheus Books, New York, pp. 55–65.

17. Midgley, M., 1985, "Evolution as a religion," Methuen Press, New York.

18. Hofstadter, R., 1944, "Social Darwinism in American Thought," Beacon Press, Boston.

19. Lambert, E.C., 1978, "Modern Medical Mistakes," Indiana Univ. Press, Bloomington.

20. Walters, L., 1986, The ethics of human gene therapy. Nature 300: 225–227.

21. Feldman, N., 1972, Camps (concentration and extermination). "Encyclopedia Judaica" 5: 77–90.

22. Beecher, H.K., 1966, Ethical and clinical research. New England Journal of Medicine 247: 1354–1360.

23. Rudner, R., 1980, The scientist qua scientist makes value judgements. In "Introductory Readings in Philosophy of Science," ed. E.D. Klemke, R. Hollinger, and A.D. Kline, Prometheus Books, New York, pp. 231–237.

24. Schutz, A., 1965, The problem of transcendental intersubjectivity in Husserl. In "Collected Papers III," ed. I. Schutz, Nijhoff Publishers, The Hague, Netherlands, pp. 51–84.

25. Whitehead, A.N., 1925, "Science and the Modern World," Macmillan, Toronto, Canada.

26. Dickson, D., 1986, Gene splicing debate heats up in Germany. Science 232: 13–14.

27. Friedlander, M.W., 1972, "The Conduct of Science," Prentice-Hall, Englewood Cliffs, N.J.

28. T.D. Lysenko, quoted in Graham, L.R., 1972, "Science and Philosophy in the Soviet Union," Alfred A. Knopf Publishers, New York.

29. Holden, C., 1986, A pivotal year for lab animal welfare. Science, 232: 147–150.

30. Garfield, The economic impact.

31. National Science Board, 1983, "Science Indicators 1982," Government Printing Office, Washington, D.C.

32. Holden, C., 1985, OTA critical of AIDS initiative. Science 227: 1182–1183.

33. Gonzalez, H.B., 1984, Scientists and congress. Science 224: 127–129.

34. National Science Board, "Science Indicators 1982."

35. Novick, J.D., 1983, Litigating the religion of creation science. Federation Proceedings 42: 3039–3042.

36. Compare with Marshall, E., 1981 Ethical risks in biomedicine. Science 212: 307–309.

37. National Science Board, "Science Indicators 1982."

38. Sun, M., 1985, TV Scientists. Science 228: 1294.

39. Goodfield, J., 1985, "Quest for the Killers," Birkauser Boston Inc., Cambridge, Mass.

40. Rothman, S., and Lichter, S.R., 1982, The nuclear energy debate: Scientists, the media, and the public. Public Opinion 5: 47–52.

41. National Science Board, "Science Indicators 1982."

42. Ibid.

Index